Essential Entomology

Essential Entomology

Alex Hall

CALLISTO
REFERENCE

www.callistoreference.com

Callisto Reference,
118-35 Queens Blvd., Suite 400,
Forest Hills, NY 11375, USA

Visit us on the World Wide Web at:
www.callistoreference.com

ISBN: 978-1-64116-561-7 (Hardback)

Cataloging-in-Publication Data

Essential entomology / Alex Hall.
    p. cm.
Includes bibliographical references and index.
ISBN 978-1-64116-561-7
1. Entomology. 2. Insects. 3. Zoology. I. Hall, Alex.
QL463 .E87 2022
595.7--dc23

Table of Contents

Preface

Entomology is a branch of zoology that focuses on the study of insects. It is a taxon based category and covers all the scientific studies which are related to insects. Some of the common insect categories which are dealt with in this field are Coleoptera, Lepidoptera, Hymenoptera and Diptera. One of its major sub-disciplines is economic entomology, which studies the harmful and beneficial effects of insects on humans and their activities. Entomology plays a vital role in the study of biodiversity and in analyzing the environmental quality. This textbook attempts to understand the multiple branches that fall under the discipline of entomology and how such concepts have practical applications. Some of the diverse topics covered herein address the varied branches that fall under this category. In this book, constant effort has been made to make the understanding of the difficult concepts of entomology as easy and informative as possible, for the readers.

A short introduction to every chapter is written below to provide an overview of the content of the book:

Chapter 1 - The branch of zoology which studies insects is known as entomology. Insects make up more than two-thirds of all the known organisms. They are primarily divided into two subclasses known as apterygota and pterygota. Apterygota are the wingless insects and pterygota are the winged insects. This is an introductory chapter which will introduce briefly all these sub-fields of entomology.; **Chapter 2** - The external anatomy of insects comprises of three main segments, namely, head, thorax and abdomen. The external body of insects also comprises of three pairs of jointed legs, antennae and compound eyes. The head of insects support a pair of sensory antennae, compound eyes and appendages. The topics elaborated in this chapter will help in gaining a better perspective about these different parts belonging to the external anatomy of insects.; **Chapter 3** - The internal anatomy of insects is made up of various organ systems such as the brain, circulatory system and excretory system. Even though there are a lot of variations between different insects, there are similarities in overall anatomy. The chapter closely examines these key components of the internal anatomy of insects to provide an extensive understanding of the subject.; **Chapter 4** - The physiological capacity of organisms that provides data for perception is known as sense. Insects are able to sense various kinds of stimuli such as light, heat, chemicals and mechanical pressure. The topics elaborated in this chapter will help in gaining a better perspective about the branches of sensory systems and behaviour in insects.; **Chapter 5** - Insects are threatened by a broad range of predators, such as birds, amphibians and carnivorous plants. The different ways in which insects defend themselves are classified as mimicry, mechanical defence and chemical defence. The chapter closely examines these key concepts of insect defence to provide an extensive understanding of the subject.

I extend my sincere thanks to the publisher for considering me worthy of this task. Finally, I thank my family for being a source of support and help.

Alex Hall

Chapter 1

Understanding Entomology

The branch of zoology which studies insects is known as entomology. Insects make up more than two-thirds of all the known organisms. They are primarily divided into two subclasses known as apterygota and pterygota. Apterygota are the wingless insects and pterygota are the winged insects. This is an introductory chapter which will introduce briefly all these sub-fields of entomology.

Insect

Insect is any member of the largest class of the phylum Arthropoda, which is itself the largest of the animal phyla. Insects have segmented bodies, jointed legs, and external skeletons (exoskeletons). Insects are distinguished from other arthropods by their body, which is divided into three major regions:

1. The head, which bears the mouthparts, eyes, and a pair of antennae.

2. The three-segmented thorax, which usually has three pairs of legs (hence "Hexapoda") in adults and usually one or two pairs of wings.

3. The many-segmented abdomen, which contains the digestive, excretory, and reproductive organs.

Paper wasp (Polistes fuscatus).

In a popular sense, "insect" usually refers to familiar pests or disease carriers, such as bedbugs, houseflies, clothes moths, Japanese beetles, aphids, mosquitoes, fleas, horseflies, and hornets, or to conspicuous groups, such as butterflies, moths, and beetles. Many insects, however, are beneficial from a human viewpoint; they pollinate plants, produce useful substances, control pest insects, act as scavengers, and serve as food for other animals (see below Importance). Furthermore, insects are valuable objects of study in elucidating many aspects of biology and ecology. Much of the scientific knowledge of genetics has been gained from fruit fly experiments and of population biology from flour beetle studies. Insects are often used in investigations of hormonal action, nerve and sense organ function, and many other physiological processes. Insects are also used as environmental quality indicators to assess water quality and soil contamination and are the basis of many studies of biodiversity.

European hornet (Vespa crabro).

Horse fly (Tabanus trimaculatus).

General Features

In numbers of species and individuals and in adaptability and wide distribution, insects are perhaps the most eminently successful group of all animals. They dominate the present-day land fauna with about 1 million described species. This represents about three-fourths of all described animal species. Entomologists estimate the actual number of living insect species could be as high as 5 million to 10 million. The orders that contain the greatest numbers of species are Coleoptera (beetles), Lepidoptera (butterflies and moths), Hymenoptera (ants, bees, wasps), and Diptera (true flies).

Eastern tailed blue butterfly (Everes comyntas; also called Cupido comyntas).

Bombardier beetle (Brachinus).

Appearance and Habits

African goliath beetle (Goliathus giganteus).

Walkingstick (Phasmatidae).

The majority of insects are small, usually less than 6 mm (0.2 inch) long, although the range in size is wide. Some of the feather-winged beetles and parasitic wasps are almost microscopic, while some tropical forms such as the hercules beetles, African goliath beetles, certain Australian stick insects, and the wingspan of the hercules moth can be as large as 27 cm (10.6 inches).

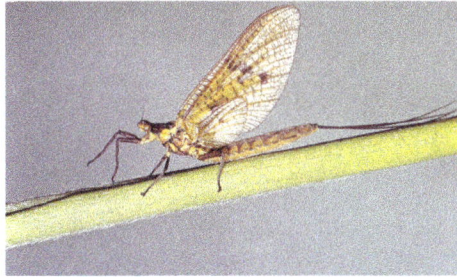

Female mayfly (Ephemera danica).

In many species the difference in body structure between the sexes is pronounced, and knowledge of one sex may give few clues to the appearance of the other sex. In some, such as the twisted-wing insects (Strepsiptera), the female is a mere inactive bag of eggs, and the winged male is one of the most active insects known. Modes of reproduction are quite diverse, and reproductive capacity is generally high. Some insects, such as the mayflies, feed only in the immature or larval stage and go without food during an extremely short adult life. Among social insects, queen termites may live for up to 50 years, whereas some adult mayflies live less than two hours.

North American firefly (Photinus).

Some insects advertise their presence to the other sex by flashing lights, and many imitate other insects in colour and form and thus avoid or minimize attack by predators that feed by day and find their prey visually, as do birds, lizards, and other insects.

Behaviour is diverse, from the almost inert parasitic forms, whose larvae lie in the nutrient bloodstreams of their hosts and feed by absorption, to dragonflies that pursue victims in the air, tiger beetles that outrun prey on land, and predaceous water beetles that outswim prey in water.

In some cases the adult insects make elaborate preparations for the young, in others the mother alone defends or feeds her young, and in still others the young are supported by complex insect societies. Some colonies of social insects, such as tropical termites and ants, may reach populations of millions of inhabitants.

Distribution and Abundance

Scientists familiar with insects realize the difficulty in attempting to estimate individual numbers of insects beyond areas of a few acres or a few square miles in extent. Figures soon become so large as to be incomprehensible. The large populations and great variety of insects are related to their small size, high rates of reproduction, and abundance of suitable food supplies. Insects abound in the tropics, both in numbers of different kinds and in numbers of individuals.

Carpenter ant (Camponotus).

If the insects (including the young and adults of all forms) are counted on a square yard (0.84 square metre) of rich moist surface soil, 500 are found easily and 2,000 are not unusual in soil samples in the north temperate zone. This amounts to roughly 4 million insects on one moist acre (0.41 hectare). In such an area only an occasional butterfly, bumblebee, or large beetle, supergiants among insects, probably would be noticed. Only a few thousand species, those that attack people's crops, herds, and products and those that carry disease, interfere with human life seriously enough to require control measures.

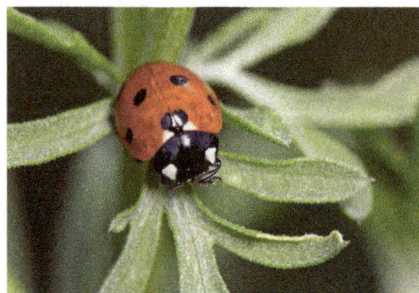

Ladybug.

Insects are adapted to every land and freshwater habitat where food is available, from deserts to jungles, from glacial fields and cold mountain streams to stagnant, lowland ponds and hot springs. Many live in brackish water up to 1/10 the salinity of seawater, a few live on the surface of seawater, and some fly larvae can live in pools of crude petroleum, where they eat other insects that fall in.

Importance

Role in Nature

Insects play many important roles in nature. They aid bacteria, fungi, and other organisms in the decomposition of organic matter and in soil formation. The decay of carrion, for example, brought about mainly by bacteria, is accelerated by the maggots of flesh flies and blowflies. The activities of these larvae, which distribute and consume bacteria, are followed by those of moths and beetles, which break down hair and feathers. Insects and flowers have evolved together. Many plants depend on insects for pollination. Some insects are predators of others.

Codling moth larva (Cydia pomonella)
parasitizing an apple.

Commercial Significance

Certain insects provide sources of commercially important products such as honey, silk, wax, dyes, or pigments, all of which can be of direct benefit to humans. Because they feed on many types of organic matter, insects can cause considerable agricultural damage. Insect pests devour crops of food or timber, either in the field or in storage, and convey infective microorganisms to crops, farm animals, and humans. The technology for combatting such pests constitutes the applied sciences of agricultural and forest entomology, stored product entomology, medical and veterinary entomology, and urban entomology.

Insects as a Source of Raw Materials

For primitive peoples who gathered food, insects were a significant food source. Grasshopper plagues, termite swarms, large palm weevil grubs, and other insects are still sources of protein in some countries. The dry scaly excreta of coccids (Homoptera) on tamarisk or larch trees is the source of manna in the Sinai Desert. Coccids were once the source of the crimson dye kermes. The cochineal, or carmine, from Dactylopius scale insects found on Mexican cacti, was used for dying cloth by the Aztecs and is used today as a dye in foods, cosmetics, drugs, and textiles. Several insect waxes are used commercially, especially beeswax and lac wax. The resinous product of the lac insect Kerria lacca (Homoptera), which is cultured for this purpose, is the source of commercial shellac.

Two of the most important domesticated insects are the silkworm (Lepidoptera) and the honeybee (Hymenoptera). Some coarse silks are produced from the cocoons of large wild silkworm species. Most commercial silks, however, come from the silkworm Bombyx mori. This insect is unknown in the wild state and exists only in culture. It was domesticated in China thousands of years ago, and selective breeding, notably in China and Japan, has produced many specialized strains. The honeybee is a close relative of existing wild bees. However, the major importance of honeybees lies in their pollination of fruit trees and other crops.

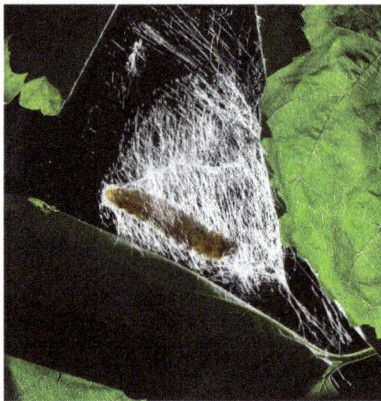

A silkworm spinning a cocoon.

Insect Damage to Commercial Products

When insects that break down dead trees invade structural timbers in buildings, they become pests. This is true of insects such as dermestid beetles and various tineid moths that ecologically are late-comers to carcasses and are capable of breaking down the keratin in hair and feathers. When these insects invade skins, furs, and wool garments or carpets, they can become problems for humans.

In many hot, dry climates, as in North Africa or the plains of India, ripened grain in the fields is invaded by certain beetles and moths. When the grain is harvested, these insects thrive in the grain stores. They can be carried throughout the world in commerce and have become universal pests of stored grain, dried fruit, tobacco, and other products. Quarantine and disinfestation methods are used to control importation of such insects from grain-exporting countries.

Agricultural Significance

Ecological Factors

Many insects are plant feeders, and, when the plants are of agricultural importance, humans are often forced to compete with these insects. Populations of insects are limited by such factors as unfavourable weather, predators and parasites, and viral, bacterial, and fungal diseases, as well as many other factors that operate to make insect populations stable. Agricultural methods that encourage the planting of ever larger areas to single crops, which provides virtually unlimited food resources, has removed some of these regulating factors and allowed the rate of population growth of insects that attack those crops to increase. This increases the probability of great infestations of certain insect pests. Many natural forests, which form similar giant monocultures, always seem to have been subject to periodic outbreaks of destructive insects.

Spotted cucumber beetle (Diabrotica undecimpunctata)

In some agricultural monocultures, nonnative insect pests have been accidentally introduced along with a crop but without also bringing along its full range of natural enemies. This has occurred in the United States with the oystershell scale (Lepidosaphes ulmi) of apple and other fruit trees, the cottony cushion scale (Icerya purchasi) of citrus, the European corn borer (Pyrausta nubilalis; also called Ostrinia nubilalis), and others. The Colorado potato beetle (Leptinotarsa decemlineata), which caused appalling destruction to the cultivated potato in the United States beginning about 1840, was a native insect of semidesert country. The beetle, which fed on the buffalo burr plant, adapted itself to a newly introduced and abundant diet of potatoes and thus escaped from all previous controlling factors. Similar situations often have been controlled by determining the major predators or parasites of an alien insect pest in its country of origin and introducing them as control agents. A classic example is the cottony cushion scale, which threatened the California citrus industry in 1886. A predatory ladybird beetle, the vedalia beetle (Rodolia cardinalis), was introduced from Australia, and within a year or two the scale insect had virtually disappeared. The success was repeated in every country where the scale insect had become established without its predators. In eastern Canada in in the early 1940s the European spruce sawfly (Gilpinia hercyniae), which had caused immense damage, was completely controlled by the spontaneous appearance of a viral disease, perhaps unknowingly introduced from Europe. This event led to increased interest in using insect diseases as potential means of managing pest populations.

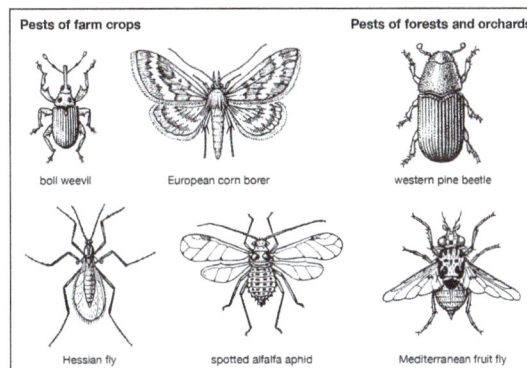

insect pests of farm crops, forests, and orchards

The boll weevil (Anthonomus grandis), European corn borer (Pyrausta nubilalis, or Ostrinia nubilalis), Hessian fly (Mayetiola destructor), and spotted alfalfa aphid (Therioaphis maculata) are important pests of farm crops. The western pine beetle (Dendroctonus brevicomis) is an insect pest found in forests, while the Mediterranean fruit fly (Ceratitis capitata) is an example of an orchard pest.

Cottony-cushion scales (Icerya purchasi, magnified).

Damage to Growing Crops

Insects are responsible for two major kinds of damage to growing crops. First is direct injury done to the plant by the feeding insect, which eats leaves or burrows in stems, fruit, or roots. There are hundreds of pest species of this type, both in larvae and adults, among orthopterans, homopterans, heteropterans, coleopterans, lepidopterans, and dipterans. The second type is indirect damage in which the insect itself does little or no harm but transmits a bacterial, viral, or fungal infection into a crop. Examples include the viral diseases of sugar beets and potatoes, carried from plant to plant by aphids.

Mexican fruit flies (Anastrepha ludens),
an invasive species, feeding on a citrus fruit.

Although most insects grow and multiply in the crop they damage, certain grasshoppers are well-known exceptions. They can exist in a relatively harmless solitary phase for a number of years, during which time their numbers may increase. They then enter a gregarious phase, forming gigantic migratory swarms, which are transported by winds or flight for hundreds or thousands of miles. These swarms may completely destroy crops in an invaded region. The desert locust (Schistocerca gregaria) and migratory locust (Locusta migratoria) are two examples of this type of life cycle.

Medical Significance

Insect damage to humans and livestock also may be direct or indirect. Direct human injury by insect stings and bites is of relatively minor importance, although swarms of biting flies and mosquitoes often make life almost intolerable, as do biting midges (sand flies) and salt-marsh mosquitoes. Persistent irritation by biting flies can cause deterioration in the health of cattle. Some blowflies, in addition to depositing their eggs in carcasses, also invade the tissue of living animals including humans, a condition known as myiasis. An example of an insect that causes this condition is the screwworm fly (Cochliomyia) of the southern United States and Central America. In many parts of the world, various blowflies infest the fleece and skin of sheep. This infestation, called sheep-strike, causes severe economic damage.

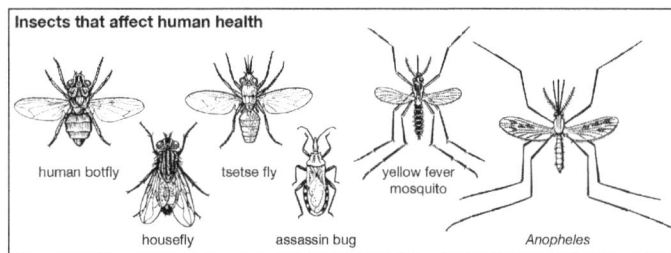

Insects that affect human health

The human botfly (Dermatobia hominis), housefly (Musca domestica), tsetse fly (genus Glossina), assassin bug (family Reduviidae), yellow fever mosquito, and Anopheles mosquito are examples of insects that are of medical significance to humans.

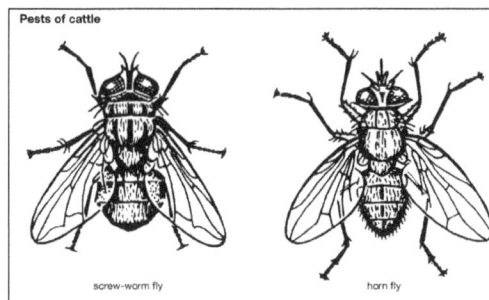

Insect pests of cattle: The screwworm and horn fly
(Haematobia irritans) are insect pests of cattle

Many major human diseases are produced by microorganisms conveyed by insects, which serve as vectors of pathogens. Malaria is caused by the protozoan Plasmodium, which spends part of its developmental cycle in Anopheles mosquitoes. Epidemic relapsing fever, caused by spirochetes, is transmitted by the louse Pediculus. Leishmaniasis, caused by the protozoan Leishmania, is carried by the sand fly Phlebotomus. Sleeping sickness in humans and a group of cattle diseases that are widespread in Africa and known as nagana are caused by protozoan trypanosomes transmitted by the bites of tsetse flies (Glossina). Under nonsanitary conditions the common housefly Musca can play an incidental role in the spread of human intestinal infections (e.g., typhoid, bacillary and amebic dysentery) by contamination of food. The tularemia bacillus can be spread by deerfly bites, the bubonic plague bacillus by fleas, and the epidemic typhus rickettsia by the louse Pediculus. Various mosquitoes spread viral diseases (e.g., several encephalitis diseases; dengue and yellow fever in humans and other animals).

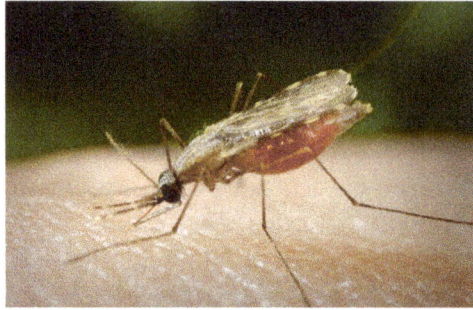

mosquito: malaria vector
Mosquito (Anopheles minimus) feeding on a human.
A. minimus is a major malaria vector in Asia.

The relationships among the various organisms are complex. Malaria, for example, has a different epidemiology in almost every country in which it occurs, with different Anopheles species responsible for its spread. These same complexities affect the spread of sleeping sickness. Some relationships are indirect. Plague, a disease of rodents transmitted by flea bites, is dangerous to humans only when heavy mortality among domestic rats forces their infected fleas to attack people, thereby causing an outbreak of plague. Typhus, tularemia, encephalitis, and yellow fever also are maintained in animal reservoirs and spread occasionally to humans.

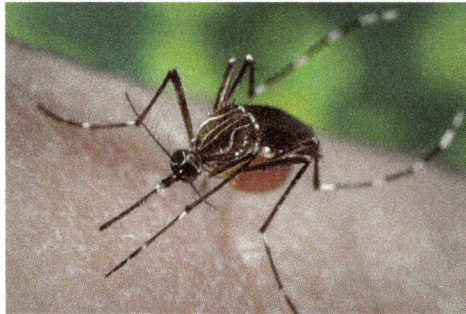

Aedes aegypti mosquito
Aedes aegypti mosquito, a carrier of yellow fever and dengue.

Control of Insect Damage

The historical objective of the entomologist was primarily to develop and introduce modifications into the environment in such ways that diseases will not be spread by insects and crops will not be damaged by them. This objective has been achieved in numerous cases. For example, in many cities flies no longer play a major role in spreading intestinal infections, and land drainage, improved housing, and insecticide use have eliminated malaria in many parts of the world.

Massive outbreaks of the Colorado potato beetle in the 1860s led to the first large-scale use of insecticides in agriculture. These highly poisonous chemicals (e.g., Paris green, lead arsenate, concentrated nicotine) were used in large quantities. The continued search for effective synthetic compounds led in the early 1940s to the production of DDT, a remarkable compound that is highly toxic to most insects, nontoxic to humans in small quantities (although cumulative effects may be severe), and long-lasting in effect. Widely used in agriculture for many years, DDT was not the perfect insecticide. It often killed parasites as effectively as the pests themselves, creating ecological imbalances that permitted new pests to develop large populations. Furthermore, resistant strains of pests appeared. The environmental longevity of many early insecticides was also found to cause significant ecological problems. Similar difficulties were encountered with many successors to DDT, such as Dieldrin and Endrin.

In the course of developing effective insecticides, the primary emphases have been to reduce their potential to cause human health problems and their impact on the environment. Biological methods of pest management have become increasingly important as the use of undesirable insecticides decreases. Biological methods include introducing pest strains that carry lethal genes, flooding an area with sterile males (as was successfully done for the control of the screwworm fly), or developing new kinds of insecticide based on modifications of insects' growth hormones. The sugar industry in Hawaii and the California citrus industry rely on biological control methods. Although these methods are not consistently effective, they are considered to be less harmful to the environment than are some chemicals.

Life Cycle

Egg

Most insects begin their lives as fertilized eggs. The chorion, or eggshell, is commonly pierced by respiratory openings that lead to an air-filled meshwork inside the shell. For some insects (e.g., cockroaches and mantids) a batch of eggs is cemented together to form an egg packet or ootheca. Insects may pass unfavourable seasons in the egg stage. Eggs of the springtail Sminthurus (Collembola) and of some grasshoppers (Orthoptera) pass summer droughts in a dry shrivelled state and resume development when moistened. Most eggs, however, retain their water although they may pass the winter in a state of arrested development, or diapause, usually at some early stage in embryonic development. However, dried eggs of Aedes mosquitoes enter a state of dormancy after development is complete and quickly hatch when placed in water.

Polyphemus moth (Antheraea polyphemus) depositing eggs. Because the larvae feed on a variety of trees and shrubs, site selection for egg deposition is haphazard.

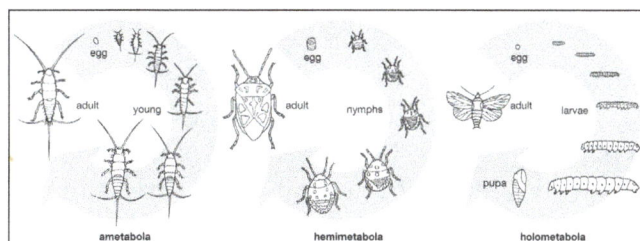

Types of insect development.

The hatching of young larvae is achieved in several ways. Some, such as caterpillars, bite their way out of the egg. Many, such as the flea, have hatching spines with which they cut a slit in the shell. Some insect eggs have a preformed "escape cap" that the larva pops from the shell by increasing the pressure inside the egg. Depending on the species, this may be accomplished either by swallowing air and then constricting muscles in the body to exert pressure on the cap or by having an expandable region on the head (many Diptera have a ptilinum) that can be extended by hydraulic (blood) pressure. After hatching, the larva continues to distend itself in this way, although the ptilinum collapses back into the body, until the cuticle hardens.

Once formed, the insect cuticle cannot grow. Growth can occur only by a series of molts (ecdyses) during which new and larger cuticles form and old cuticles are shed. Molting makes possible large changes in body form.

Types of Metamorphosis

In the most primitive wingless insects (apterygotes) such as the silverfish Lepisma saccharina, there is almost no change in form throughout growth to the adult. These are known as ametabolous insects. Among insects such as grasshoppers (Orthoptera), true bugs (Heteroptera), and homopterans (e.g., aphids, scale insects), the general form is constant until the final molt, when the larva undergoes substantial changes in body form to become a winged adult with fully developed genitalia. These insects, called hemimetabolous, are said to undergo incomplete metamorphosis. The higher orders of insects, including Lepidoptera (butterflies and moths), Coleoptera (beetles), Hymenoptera (ants, wasps, and bees), Diptera (true flies), and several others, are called holometabolous because larvae are totally unlike adults. These larvae undergo a series of molts with little change in form before they enter into complete metamorphosis, which includes molting first into pupae and then into fully winged adults.

A molting insect shedding its exoskeleton.

Types of Larvae

Larvae, which vary considerably in shape, are classified in five forms: eruciform (caterpillar-like), scarabaeiform (grublike), campodeiform (elongated, flattened, and active), elateriform (wireworm-like), and vermiform (maggot-like). The three types of pupae are: obtect, with appendages more or less glued to the body; exarate, with the appendages free and not glued to the body; and coarctate, which is essentially exarate but remaining covered by the cast skins (exuviae) of the next to the last larval instar (name given to the form of an insect between molts).

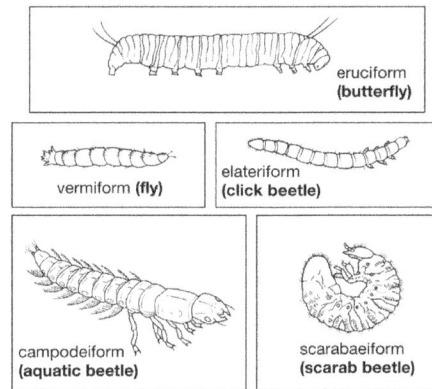

Insect larvae.

Role of Hormones

Both molting and metamorphosis are controlled by hormones. Molting is initiated when sensory receptors in the body wall detect that the internal soft tissues have filled the old exoskeleton and trigger production of a hormone from neurosecretory cells in the brain. This hormone acts upon the prothoracic gland, an endocrine gland in the prothorax, which in turn secretes the molting hormone, a steroid known as ecdysone. Molting hormone then acts on the epidermis, stimulating growth and cuticle formation. Metamorphosis likewise is controlled by a hormone. Throughout the young larval stages a small gland behind the brain, called the corpus allatum, secretes juvenile hormone (also known as neotenin). As long as this hormone is present in the blood the molting epidermal cells lay down a larval cuticle. In the last larval stage, juvenile hormone is no longer produced, and the insect undergoes metamorphosis into an adult. Among holometabolous insects the pupa develops in the presence of a very small amount of juvenile hormone.

Although a state of arrested development may occur during any stage, diapause occurs most commonly in pupae. In temperate latitudes many insects overwinter in the pupal stage (e.g., cocoons). The immediate cause of diapause, failure to secrete the growth and molting hormones, usually is induced by a decrease in daylength as summer wanes.

In addition to changes in form during development, many insects exhibit polymorphism as adults. For example, the worker and reproductive castes in ants and bees may be different, termites have a soldier caste as well as reproductives and persistent larvae, adult aphids (Homoptera) may be winged or wingless, and some butterflies show striking seasonal or sexual dimorphism. The general interpretation of all such differences is that, although the capacity to develop different forms is present in the genes of every member of a given species, particular lines of development are evoked by environmental stimuli. Hormones, including perhaps juvenile hormone, may be agents for the control of such changes.

Reproduction

The life of the adult insect is geared primarily to reproduction. Since reproduction is sexual in almost all insects, mating must be followed by impregnation of the female and fertilization of eggs. Usually the male seeks out the female. In butterflies in which vision is important, the colour of the female in flight can attract a male of the same species. In mayflies (Ephemeroptera) and certain midges (Diptera), males dance in swarms to provide a visual attraction for females. In certain beetles (e.g., fireflies and glowworms) parts of the fat body in the female have become modified to form a luminous organ that attracts the male. Male crickets and grasshoppers attract females by their chirping songs, and the male mosquito is lured by the sound emitted by the female in flight. The most important element in mating, however, is odour. Most female insects secrete odorous substances called pheromones that serve as specific attractants and excitants for males. The male likewise may produce scents that excite the female. Certain scales (androconia) on the wings of many male butterflies function in this way. Assembling scents, active in small quantities, are well known in female gypsy moths and silkworms as male attractants. The queen substance in the honeybee serves the same purpose.

Silkworm moths mating

Mating and egg production require appropriate temperatures and adequate nutrition. The need for protein is particularly important, and in insects such as Lepidoptera (butterflies and moths), which take only sugar and water in the adult stage, necessary protein is derived from larval reserves. Temperature and nutrition often influence hormone secretion. Juvenile hormone or hormones from the neurosecretory cells commonly are needed for egg production. In the absence of these hormones reproduction is arrested, and the insect enters a reproductive diapause. This phenomenon occurs in potato beetles of genus Leptinotarsa during the winter.

A few insects (e.g., the stick insect Carausius) rarely produce males, and the eggs develop without fertilization in a process known as parthenogenesis. During summer months in temperate latitudes, aphids occur only as parthenogenetic females in which embryos develop within the mother (viviparity). In certain gall midges (Diptera) oocytes start developing parthenogenetically in the ovaries of the larvae, and the young larvae escape by destroying the body of their mother in a process called paedogenesis.

Evolution and Paleontology

Origin of Insects

The most primitive insects known are found as fossils in rocks of the Middle Devonian Period (393.3 million to 382.7 million years ago). The bodies of those insects were divided then, as now, into a head bearing one pair of antennae, a thorax with three pairs of legs, and a segmented abdomen. Those insects originated with the terrestrial branch of the phylum Arthropoda. The Arthropoda, whose origin is thus far unknown, probably arose in Precambrian times, perhaps as many as 1 billion years ago. Some arthropods colonized the open sea and have become the present-day class Crustacea (crabs, shrimps) and the now-extinct Trilobita. Other arthropods colonized the land.

This terrestrial line persists chiefly as the classes Onychophora, Arachnida (spiders, scorpions, ticks), the myriapods (consisting of Diplopoda [millipedes], Pauropoda, Symphyla, and Chilopoda, or centipedes), and finally the class Insecta.

The most primitive insects today are found among the wingless (apterous) hexapods; sometimes known collectively as apterygotes, they include proturans, thysanurans, diplurans, and collembolans. It is agreed generally that insects are related most closely to the myriapod group, among which the Symphyla exhibit most of the essential features required for the ancestral insect form (i.e., a Y-shaped epicranial suture, two pairs of maxillae, a single pair of antennae, styli and sacs on the abdominal segments, cerci, and malpighian tubules). There is, therefore, general agreement that the insects probably arose from an early symphylan-like form.

Insect Fossil Record

The insect fossil record has many gaps. Among the primitive apterygotes, only the collembolans (springtails) have been found as fossils in the Devonian Period (about 419.2 million to 358.9 million years ago). Ten insect orders are known as fossils, mostly of Late Carboniferous and Permian times (318 million to 251 million years ago). No fossils have yet been found from the Late Devonian (about 382.7 million to 358.9 million years ago) or Early Carboniferous (about 358.9 million to 318 million years ago) period, when the key characters of present-day insects are believed to have evolved; thus, early evolution must be inferred from the morphology of extant insects.

It has become evident that insect evolution, like that of other animals, was far more active at some periods than at others. There have been geological epochs of "explosive" evolution during which many new forms have appeared. Those epochs may have followed some modification or innovation in body function, or new developments favoured by climatic changes or evolutionary advances of other animals and plants. During those periods of evolutionary change, new methods of feeding and living led to diversity of insect mouthparts and limbs, the origin of metamorphosis, and other changes.

Insect Phylogeny

The simplified family tree shown here illustrates the presumed evolutionary history of winged insects (Pterygota) throughout the geological periods from the Devonian to the Recent. The apterygotes, which are regarded as survivors of primitive insect stock, are omitted from the family tree. Dark lines indicate the periods during which the various orders have been found as fossils. Some lines stop at the names of orders now extinct and known only as fossils. Light lines indicate the hypothetical origin of various orders. Many insect types, traces of which have not yet been discovered, must have been produced during the explosive periods of evolution in Carboniferous and Permian times.

The primitive wingless insects gave rise to a paleopterous stock. Descendants of this stock included ancient fossil types that flourished in Permian times, such as the giant dragonflies or Protodonata (some of which had a wing span of more than half a metre) and the dragonflies and damselflies (Odonata) and mayflies (Ephemeroptera), both of which have persisted with little change to the present. The primitive insect stock also gave rise to a neopterous stock, believed to include the progenitors of the remaining insect orders. The Orthoptera (grasshoppers) and the Plecoptera (stoneflies) have been found as fossils even in late Carboniferous times. The Isoptera (termites, sometimes placed in the order Blattodea), Embioptera (webspinners), and Dermaptera (earwigs), though doubtless of ancient origin, have not been found yet as fossils dated earlier than the Mesozoic Era (252 million to 66 million years ago).

The evolutionary radiation believed to have given rise to the orders listed above in the Middle Carboniferous Period is thought to have also produced a paraneopterous stock, which formed the base for a new evolutionary radiation during the Permian Period. Present-day derivatives of this stock evolved into the Psocoptera (psocids), Mallophaga (chewing lice), Anoplura or Siphunculata (sucking lice), Thysanoptera (thrips), Heteroptera (true bugs), and Homoptera (e.g., aphids).

Several phylogenetic lines are exopterygote—i.e., insects with simple metamorphosis, some of which, such as Mallophaga and Anoplura, are secondarily wingless. The remaining orders are endopterygote (insects with complete metamorphosis). They are shown in the family tree as derivatives of an oligoneopterous stock, which gave rise to Neuroptera (lacewings), Hymenoptera (ants, wasps, and bees), and Coleoptera (beetles) in the Early Permian Period (298.9 million to 272.3 million years ago); the early ancestry of these orders is obscure, however, and the earliest fossils closely resemble present-day forms. One line from the evolutionary radiation at the beginning of the Permian gave rise to a mecopteroid stock, and there is good evidence that a sub-radiation of these mecopteroid orders (sometimes called the panorpoid complex) provided the origin for the present Mecoptera (scorpionflies), Diptera (true flies), Siphonaptera (fleas), Trichoptera (caddisflies), and Lepidoptera (butterflies and moths).

Caddisfly (order Trichoptera).

Evolution

Wings and Flight

Insect wings develop as paired outgrowths from the thorax, stiffened by ribs, or veins, in which run tracheae. These tracheae follow a consistent pattern throughout the Pterygota, and their specific modifications (known as venation) are important in classification and in estimations of the degree of relationship between groups. The basic consistency of venation suggests that wings have been evolved only once among the insects; that is, all the Pterygota arose from a single stem in the family tree. By the time fossil insects are found (toward the end of the Carboniferous), wings are developed fully. In the Paleoptera the wings are held aloft above the back, as in mayflies, or held extended permanently on each side of the body, as in dragonflies. Throughout the Neoptera there is a wing-flexing mechanism (secondarily lost in butterflies) that enables the wings to be folded back to rest on the surface of the abdomen.

Winged insects must have made their appearance very early in the Carboniferous, but there is no fossil evidence to show the way they evolved. One hypothesis is that wings arose as fixed planes extending sideways from the thorax and that these planes were used, perhaps in some large leaping insect, for gliding. Later muscles developed, first to control inclination and then to move the wings in flapping flight. Another hypothesis is that wings may have originated from large thoracic tracheal gills, similar to the movable tracheal gills along the abdomen of some mayfly larvae. Such outgrowths could have been useful to insects exposed by the drying up of a temporary aquatic habitat and might have carried them in rain-bearing winds to a new watery home. It is likely that the most primitive symphylan-like insects were terrestrial. Throughout insect evolution, however, independent adaptations to aquatic habitats have occurred. Usually the pattern is one in which the adults leave the water and disperse. Many pterygote insects have become secondarily wingless, sometimes as single species or groups of species within large orders and sometimes as entire orders (the parasitic lice, Mallophaga and Anoplura, and the fleas, Siphonaptera).

Metamorphosis

It generally is agreed that insect metamorphosis evolved as adult insects gradually adopted different modes of life from those of larvae. The characters of larva and adult became genetically independent; in response to natural selection, therefore, each was able to evolve independently of the other. Mouthparts, limbs, and other morphological features were modified in different directions and in higher groups. Where these differences were extreme, an intermediate pupal stage evolved to bridge the morphological gap between larva and adult. It seems quite probable that the development of metamorphosis occurred more than once during the evolution of insects.

Feeding Methods

Insects did not evolve in a constant environment. Throughout geological time there were prodigious changes in climate; in addition, evolution was continuous among all other animals and plants. Geologically the selection pressures among insects were changing continuously. At the end of the Mesozoic Era the first flowering plants appeared. Insect evolution has paralleled that of the flowering plants; they have evolved together. As Lepidoptera (butterflies and moths), Hymenoptera (ants, bees, and wasps), Diptera (true flies), and Coleoptera (beetles) began to feed upon flowers, nectar, or pollen, flowering plants came to rely more and more upon insects—rather than upon the wind—for transferring their pollen. Flowers evolved nectaries, scents, and conspicuous colours as attractants for those insects that could effect cross-pollination. Insects likewise evolved appropriate mouthpart modifications for extracting nectar from flowers.

During the Mesozoic warm-blooded animals (mammals and birds) first appeared; by the dawn of the Paleogene Period, they had become predominant among Earth's large animals. The warm fermenting excrement and the decaying dead bodies of mammals furnished excellent nutrient media for many insect larvae, notably among the Diptera and Coleoptera. The adults in both groups found their nourishment in flowers. Some heteropterans (true bugs) and dipterans pierce the skin of birds and mammals and feed on their blood. The Anoplura (sucking lice) and the Siphonaptera (fleas) have become so specialized for this type of parasitic existence that their relationships to other insects are not yet known with certainty.

Continuing Evolution

Evolution is occurring among present-day insects. They exhibit a balanced genetic polymorphism; in other words, in response to small environmental changes, one genetic form, more successful than another, will become more plentiful. Sometimes there is no visible difference between these forms, the advantage presumably lying in some physiological change. It is advantageous for a species to have a gene pool from which favourable characters can be selected so that the species can respond to environmental changes. Changes within a species may occur progressively over a large geographical area. Such a progressive genetic change is called a cline; in some cases insects at the extremes of the cline are so unlike that they are taken as separate species and may be infertile when crossed.

One well-known example of evolution in action among insects is industrial melanism (accumulation of the black pigment melanin); many butterflies inhabiting industrial areas have become almost black, the black forms being more tolerant of pollution and less conspicuous to predators. Another example of this cline type of evolution is the development of insect strains resistant to an insecticide that has been applied heavily in an area for several years. In many parts of the world houseflies became highly resistant to DDT.

Insect Taxonomy

Taxonomy is the science of classifying organisms. Taxonomy results in classifications, which allow for storage, retrieval and communication of information about organisms. A key function of taxonomy is to provide correct identification of organisms. This workshop introduces participants to the taxonomic process, how to accurately identify insects to order, and what steps can be taken to identify insects to a finer taxonomic level.

The methods of taxonomy include: the discovery of species, the recognition and diagnosing of taxa on the basis of characters (e.g., morphological, molecular, behavioural, etc), the formal description and naming of species, and the placement of species within a hierarchical classification.

The purpose of classifications is to order organisms on Earth into a stable and universal system that enables scientists and other members of society to communicate about them. Biological classifications have high information content which allow us to store information about a taxon's morphology, genetics, distribution, hosts, ecology and life cycles.

The Class Insecta is placed in the subphylum Hexapoda in the Phylum Arthropoda, which includes five subphyla:

- Pycnogonida (sea spiders, marine).

- Euchelicerata (spiders, mites, scorpions, ticks, harvestmen, horseshoe crabs, solifugids).

- Myriapoda (centipedes, millipedes, symphylans, pauropods).

- Crustacea (crabs, lobsters, ostracods, amphipods, shrimp, malacastrocans, barnacles).

- Hexapoda (insects, springtails, proturans, diplurans).

Arthropods (Phylum Arthropoda) are defined morphologically by the following characters:

- Bilaterally symmetrical body.

- Rigid exoskeleton.

- Jointed limbs.

Hexapods (subphylum Hexapoda) are composed of two classes, the Entognatha and the Insecta.

They are recognised by the following characters:

- Body divided into head, thorax and abdomen (tagma).

- One pair of antennae.

- Three pairs of mouthparts.

- Three pairs of uniramous (unbranched) limbs.

- Tracheal system of respiration.

The Entognatha have enclosed mouthparts and includes three orders, the springtails (Collembola), proturans (Protura) and diplurans (Diplura).

The Insecta comprise 27-30 orders of insects, depending on the classification used. For example, some authors maintain the termites as a separate order (Isoptera), but modern classification has them nested within the order Blattodea, which includes the roaches.

There are 25 orders of insects known from Papua New Guinea. The orders not currently known from Papua New Guinea include the Raphidioptera, Mecoptera, Megaloptera, Grylloblattodea, and Mantophasmatodea. The Zoraptera and Plecoptera are each represented in Papua New Guinea by a single species.

Apterygote

Apterygote is any of the primitive wingless insects, distinct from the pterygotes, or winged insects. Used in this sense, the term apterygote commonly includes the primitive insects of the following groups: proturans, collembolans (springtails), diplurans, and species in the orders Zygentoma, Archaeognatha, and Monura (formerly the thysanurans).

The taxonomic status of the various groups that are considered apterygotes, however, remains unsettled. A typical apterygote, for example, is wingless and has six legs. The presence of six legs was once an important feature in the identification of true insects and enabled the identification of the apterygotes, including the proturans, collembolans, and diplurans and the now defunct Thysanura—the four groups that together made up the traditional subclass Apterygota (class Insecta). However, the proturans, collembolans, and diplurans are now considered by some entomologists to be offshoots from the main insectan stem of evolution and have been given independent tax-onomic status as classes equivalent to the class Insecta. The term apterygote, therefore, is some-times applied only to those groups thought to be ancestors of pterygotes—i.e., the silverfish, fish-moths, and firebrats (order Zygentoma) and the bristletails (order Archaeognatha), together with the extinct monurans (order Monura).

A proturan.

General Features

Protura are minute (to 2 mm [0.08 inch] in length), elongated, and white and lack antennae. Dis-tributed throughout the world in soil and leaf litter, they number about 800 species. Collembola are diverse in form, coloration, and habitat. Most species are less than 3 mm (0.1 inch) in length, but some range to 10 mm (0.4 inch). They have either elongated or globular bodies with anten-nae and may have a furcula (ventral abdominal springing organ). Collembolans occur in soil and leaf litter throughout the world, including Antarctica. There are more than 8,200 known species. Diplura are white or yellowish, blind, elongated with long antennae, and less than 10 mm in length, although one group attains 50 mm (2 inches). Their two tail filaments (or threadlike structures) can be long and thin, short and thick, or in the form of pincers. Diplurans are widely distributed in soil, leaf litter, and rotting logs. More than 800 species have been described. Zygentoma and Archaeognatha are mainly long with three elongated tail filaments. Mostly 5 to 20 mm (0.2 to 0.8 inch) in length when fully grown, these insects are widely distributed in leaf litter, although some live in ant and termite nests. Zygentoma have small compound eyes and styli (i.e., bristlelike pro-cesses) only on the abdomen. Archaeognatha have large compound eyes and styli on the legs and abdomen. More than 350 species of Zygentoma and Archaeognatha are known. The largest of the extinct Monura were about 30 mm (1.2 inches) in length.

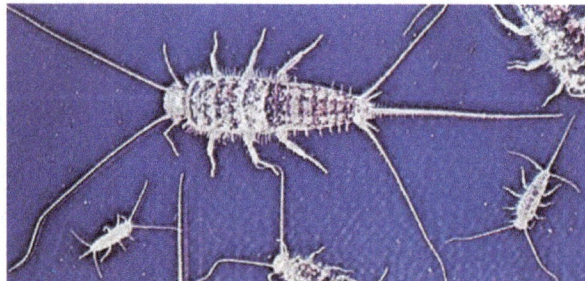

Firebrat (Thermobia domestica).

Life Cycle

The immature stages in all apterygotes are called nymphs. The young are similar to adults, changing little (slight metamorphosis) from molt to molt until sexual maturity is attained. In some groups, molting may continue throughout adult life. The greatest changes occur in the Protura, which is the only anamorphic hexapod group (i.e., an increase in number of body segments occurs at time of molting). The complete number of segments is present only after the third molt. There are at least six stages between molts (instars), and the last is the adult. Little is known about the postembryonic development of Diplura. Most species feed on both living and dead vegetable matter and fungi, although one group preys on other small invertebrates.

Springtail

The life cycles of the Collembola are diverse. Females may lay up to 800 eggs that hatch in 2 to 40 days. Three to 12 juvenile molts occur in intervals ranging from 11 days to a year, with up to 50 molts occurring in a lifetime, which can last from 4 to 18 months. Most Collembola feed on living or decaying plant material as well as on fungi, algae, and spores, while a few feed on carrion or are predatory.

In Zygentoma there may be more than 40 molts, although the adult stage is usually reached after about 12 molts. The silverfish (Lepisma saccharina) reaches sexual maturity in two or three years and molts multiple times in each subsequent year (sometimes molting more than two dozen times in a single year). They can live as long as seven years. In Archaeognatha there are six instars including adults. Both Zygentoma and Archaeognatha feed on decaying or dried vegetable material. Domestic silverfish eat plant and animal remains, paper, and artificial silk.

Silverfish (Lepisma saccharina).

Reproduction in apterygote groups is mainly sexual, but parthenogenesis (reproduction without fertilization) can occur. Males deposit sperm packets, often haphazardly, that are taken up by females and stored until the time for fertilization. In proturans small external genitalia of unusual form surround the gonopore in both sexes. Collembolans lack external genitalia, but the gonopore and surrounding area differ. In diplurans external genitalia are absent or vestigial. Zygentoma and Archaeognatha have external genitalia similar to those of the pterygotes. However, the aedeagus in males is used to deposit sperm drops and not as a copulative organ. The deposition and pickup of sperm drops in Zygentoma and Archaeognatha must take place during each adult stage if young are to be produced since the contents of the female sperm receptacle are lost with each molt.

Ecology

Most apterygotes live in soil and leaf litter or are associated with plants and rotting logs in moist regions. Collembolans are also found in aquatic environments, caves, permanent ice fields and snow, and insect and animal nests. Moisture is an important ecological factor in habitat selection. Most primitive wingless hexapods have a thin exoskeleton and must guard against dehydration.

The condition and nature of soil are important factors in the distribution and nature of collembolan populations. Fungi in soil are especially important as collembolan food. Physical factors known to affect collembolans often do so secondarily, with the primary effect being on the soil microflora. Changes in macroflora also affect collembolan populations. Direct cultivation (e.g., plowing) has an immediate harmful effect on collembolan populations but does little long-range damage. Fertilizers increase the numbers of collembolans in soil, and herbicides have no effect. Some insecticides are lethal, while others are not. Increases in soil collembolans following insecticide applications are probably due to lethal effects on predators. Estimates of soil collembolans average about 100,000 per cubic metre. Soil collembolans usually form aggregations. Certain species are more prevalent in a given layer (e.g., leaf litter, the fermentation layer, the humus-forming layer, or deep in the soil). Typically small species and young insects occur in deeper layers, although daily and seasonal vertical movements may occur. Occasional swarming of large numbers of collembolans occurs after unusually wet conditions.

The most important collembolan predators are mites. Others are pseudoscorpions, staphylinid beetles and carabid beetles, an empid fly, and dacytine ants. Occasional collembolan feeders include spiders, fish, frogs, miscellaneous ants, and pulmonate snails. Collembolans are parasitized by gregarines, nematodes, viruses, bacteria, fungi, and horsehair worms.

Form and Function

The bodies of proturans, collembolans, diplurans, Zygentoma, and Archaeognatha are divided into three segments: head, thorax, and abdomen. The three-segmented, leg-bearing thorax has one pair of legs on each segment. The extinct monurans, considered an offshoot of the Zygentoma-Archaeognatha stem, have no separate abdomen or thorax; three pairs of legs occur on the first 3 of 14 segments. The monuran body, although similar to present-day Zygentoma, has one tail filament or telson (the median tail filament in present-day Zygentoma-Archaeognatha and winged mayflies). Many proturans and collembolans have entirely cutaneous respiration (breathing through the skin). In some proturans, collembolans, diplurans, and archaeognathans, a breathing system of tracheae and spiracles (external openings) occurs in the trunk, while in Zygentoma the tracheal system is like that in pterygote insects.

Head

Proturans have anteriorly directed, prognathous mouthparts and a pair of sense organs. Collembola have either prognathous or hypognathous, ventrally directed, mouthparts and may or may not have a pair of postantennal sense organs and lateral ocelli (simple eyes). The diplurans lack eyes and sense organs, while the Zygentoma have simple lateral eyes and are hypognathous. Archaeognatha have both large compound eyes and ocelli. Monurans had large compound eyes. Mouthparts are important taxonomically in primitive wingless hexapods. Movable mouthparts are generally recognized as walking limbs that were attached to ancestral body segments that became fused to form the posterior part of the head. The Protura, Collembola, and Diplura are entognathous, meaning that the mouthparts are withdrawn inside the head capsule. Monura, Zygentoma, and Archaeognatha are ectognathous, meaning that the mouthparts are exposed, as in pterygote insects. Mandibles are present in the biting mouthparts of Diplura, Zygentoma, Archaeognatha, Monura, and most Collembola. Although proturan mouthparts are modified for sucking, mandibles are retained. However, in collembolans with sucking mouthparts, mandibles may or may not be retained.

Body

In general the three segments of the thorax are distinct, and each bears one pair of legs. Each leg has a terminal claw. Proturan legs are simple except that the anterior legs are longer, serving as tactile organs with the anterior tarsus having sensory organs. The three pairs of legs are similar to each other in the apterygote groups.

The abdomen of Protura undergoes anamorphosis: in the first and second instars it has 9 segments, the third 10, and the rest 12. Collembola have a maximum of six abdominal segments. Diplura, Zygentoma, Archaeognatha, and the extinct Monura have 11 abdominal segments. The final abdominal tailpiece is the telson.

Modified Abdominal Structures

Primitive hexapods have abdominal structures that represent modified remnants of ancestral walking limbs. Many hexapods have cerci (sensory appendages) on the 11th abdominal segment, which aid in identification of the telson. The Protura, Collembola, and Monura lack cerci. In Diplura a pair of cerci arise from the small terminal segment. Cerci can be long with numerous segments, short with a central duct and terminal pore, or modified into a pair of pincers for holding prey. They are important taxonomic criteria in separating hexapod groups. The extinct monurans had pairs of abdominal structures, called ventral styli, that are modified limbs. Similar styli are present in Zygentoma and Archaeognatha. These hexapods also have one or two pairs of vesicles on some abdominal segments, as do proturans. Collembolans have a single tube containing a pair of vesicles, a single median tenaculum (catchlike process that holds furcula in place), and a single median furcula (springing organ). Each of these is a modified pair of ancestral limbs.

Locomotion

The primitive wingless hexapods are walking terrestrial invertebrates. Proturans, however, use only the middle and posterior legs for locomotion. The anterior pair of legs are held above the head and used as tactile organs. Collembolans have a furcula, a ventral abdominal springing organ that enables them to spring rapidly into the air. Apparently, Zygentoma and Archaeognatha have developed the median tail filament (telson) to assist in supporting the abdomen during movement. Archaeognathans can also spring with considerable agility by using the body and tail filaments.

Bristletails

Bristletails are about 300-400 species of small, elongate, terrestrial insects in the order Thysanura. Bristletails have an ancient evolutionary lineage, and they are believed to be relatively primitive, that is, similar in form and function to the most early evolved insects.

Bristletails have a simple metamorphosis, with three life stages: egg, nymph, and adult. Both the nymphal and adult stages are wingless, and they are rather similar in their physical appearance, the main difference being in size and sexual maturity. The adults are large and mature.

Bristletails are easily distinguished by the three threadlike appendages that emerge from the end of their abdomen, by their long, backward-pointing antennae, and by their body covering of glistening scales. Bristletails have chewing mouth parts, and they feed on a wide range of types of soft, usually decaying organic matter. Bristletails hide during the day, and if they are disturbed they run quickly away to seek a new hiding place. Otherwise, bristletails are active at night. Bristletails are unusual in that they continue to grow and molt even after they have become sexually mature adults. Almost all other insects stop growing after they become breeding adults.

The most familiar bristletails are those in the family Lepismatidae, including the light-colored common silverfish (Lepisma saccharina) and the darker-colored firebrat (Thermobia domestica), which often occurs in warm, moist places, for example, near hearths, stoves, and furnaces. Both of these species can be common in moist places in buildings, where they feed on a wide variety of starchy foods, and can sometimes cause significant damages to books, wallpaper, fabrics, and stored foods.

Jumping Bristletails

Jumping bristletails are in the order Archaeognatha. The bristletail's "ancient jaw" has only one condyle (a bone that looks like a knuckle) attaching it to the head. This is distinctive because insects of a higher order have two condyles.

Fossil evidence shows bristletails are among the most primitive of living insects and first appeared along with arachnids in the mid-Devonian period, which was 419.2 to 358.9 million years ago. Jumping bristletails are found around the world, including the Arctic. There are about 250 species in the world, 24 of them in North America.

Physical Description

These insects are called "jumping" bristletails because they can launch themselves several inches through the air. They accomplish this by thrusting their abdominal muscles rapidly and forcefully against the ground, providing them lift-off.

Jumping bristletails have a resemblance to silverfish and firebrats and were classified with them at one time. They're similar to silverfish, in that they're wingless, scurry around much the same way and are fast runners, but, there are differences.

Silverfish can't launch themselves as the bristletails do. The bristletails don't reproduce indoors and seldom inhabit houses, as silverfish do. The bristletails are brownish or yellowish and have mottled or nondescript patterning, whereas silverfish are aptly named for their silver body. Another noticeable difference is the body shape — the bristletails have an arched thorax (the area between their head and abdomen), while silverfish are flattened.

And, finally, the eyes of jumping bristletails are large and touch each other in the middle, whereas a silverfish's eyes are smaller and don't touch.

The "bristles" in the bristletail's name refer to many small, moveable appendages under the abdomen, which give the appearance of bristles. They're called style (plural: styli) and scientists speculate that eons ago they may have been limbs. Jumping bristletails have three long appendages projecting from the tip of their abdomen, with the middle one going straight back and being much longer than the outer two.

Notice how the eyes touch in the middle.

Life Cycle

Jumping bristletails undergo simple metamorphosis, progressing from egg to nymph to adult. There are both male and female bristletails, but the two don't get together in the usual sense. Well, actually, they don't get together at all — the male merely leaves a packet of sperm (a spermatophore) on the ground for a female to find! It seems a rather happenstance system, doesn't it? Males of some species perform courtship rituals to ensure that females find their spermatophore.

Females find a crevice in which to lay their eggs. The eggs may stay dormant for a year before hatching. Nymphs are tiny versions of the parents, except they don't have scales. Over time, they'll undergo at least seven molts, as they grow larger and larger and outgrow their skin. After reaching their full adult size, which may take up to two years, they'll continue to periodically molt and may live another year or two.

Silverfish

The silverfish (Lepisma saccharina) is a small insect pest found around the globe. They are considered a nuisance pest, meaning that they are neither harmful to humans nor spread disease. Instead, silverfish are known for damaging material goods, such as books, wallpaper, photos, clothing, and dry foods in the kitchen pantry.

Silverfish are particularly attracted to under-sink areas in the bathroom or kitchen, where the environment is humid and dark. They may also live in walls, closets, or crawl spaces. They remain hidden during the day, but at night, they emerge to forage for food.

A similar insect pest, the firebrat, looks and behaves much like the silverfish. Both pests can be controlled using the same techniques.

Identification

Silverfish

Silverfish are small insects with long, thin, carrot-shaped bodies. They typically reach no more than three quarters of an inch in length and have a silvery-gray coloration. Their bodies have no obvious segmentation, and adults are covered with thin scales resembling those of a fish. They are easily identified by the two long antennae stemming from their heads and the three tail-like appendages on their back-ends. They are smaller and thinner than cockroaches, and a different color than the similarly-sized earwig.

Silverfish have rather stubby legs, but don't let that fool you. They are capable of moving very, very quickly, especially when startled. This is when most people see silverfish: when the pests' late-night feasting is interrupted and they scurry back to the dark corners of your home. Their rapid side-to-side movement resembles a swimming fish, which is said to be the source of their name.

Pterygota

Pterygota, treated as an infraclass, are the winged or secondarily wingless (apterous) insects, with thoracic segments of adults usually large and with the meso- and metathorax variably united to form a pterothorax. The lateral regions of the thorax are well developed. Abdominal segments number 11 or fewer, and lack styles and vesicular appendages like those of apterygotes. Most Ephemeroptera have a median terminal filament. The spiracles primarily have a muscular closing apparatus. Mating is by copulation. Metamorphosis is hemi- to holometabolous, with no adult ecdysis, except for the subimago (subadult) stage in Ephemeroptera.

Informal Grouping "Palaeoptera"

Insect wings that cannot be folded against the body at rest, because articulation is via axillary plates that are fused with veins, have been termed "palaeopteran" (old wings). Living orders with such wings typically have triadic veins (paired main veins with intercalated longitudinal veins of opposite convexity/concavity to the adjacent main veins) and a network of cross-veins. This wing venation and articulation, together with paleontological studies of similar features, was taken to imply that Odonata and Ephemeroptera form a monophyletic group, termed Palaeoptera. The group was argued to be sister to Neoptera which comprises all remaining extant and primarily winged orders. However, reassessment of morphology of extant early-branching lineages and recent nucleotide sequence evidence fails to provide strong support for monophyly of Palaeoptera. Here we treat Ephemeroptera as sister group to Odonata + Neoptera, giving a higher classification of Pterygota into three divisions.

Mayfly

Mayfly is any member of a group of insects known for their extremely short life spans and emergence in large numbers in the summer months. Other common names for the winged stages are shadfly, sandfly, dayfly, fishfly, and drake. The aquatic immature stage, called a nymph or naiad, is widely distributed in freshwater, although a few species can tolerate the brackish water of marine estuaries.

The winged stages attract attention through mass emergences when they may make roads slippery, clog gutters, and taint the air with an odour of decay. Mayfly nymphs are important in the energy transfer cycle that occurs in freshwaters. Some species are carnivorous, but the majority of nymphs feed on diatoms, algae, higher plants, and organic detritus. Nymphs are devoured in turn by many carnivorous animals, especially fishes.

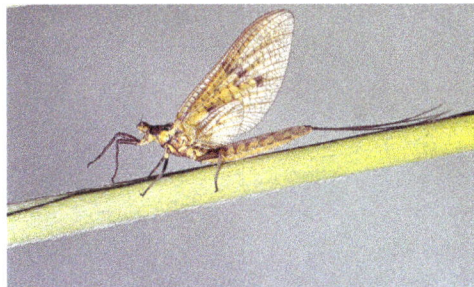

Female mayfly (Ephemera danica).

Features

Appearance

Winged mayflies have large compound eyes, short, bristlelike antennae, and functionless mouthparts and digestive tracts. Once mayflies enter the winged stages they cannot feed. Their membranous wings include a large, triangular front pair and a much smaller, rounded hind pair. In a few species, the hind pair is extremely reduced or absent. In repose, the wings are held together upright over the body like those of a butterfly. The adult mayfly has two or three threadlike tails, usually as long as, or longer than, the body.

Nymphal characters include a single claw terminating each of the six legs. The surface of the thoracic region of the body is strongly rounded outward and bears the developing wings in external pads on the upper surface. The abdominal region is usually long and slender. Gills are attached to the outer edge of the upper surface of some of the ten segments into which the body is divided. The body of the nymph terminates in three, less often two, slender tails. Adult mayflies of North American species range in body length, exclusive of tails, from 2.5 mm (0.1 inch) for Caenis to 32 mm (more than an inch) for Hexagenia.

Distribution and Abundance

Worldwide, about 2,500 species of mayflies have been described, about 700 of them from North America north of Mexico. The order is represented on all continents except Antarctica. In areas of high biological productivity (e.g., gravel-bottomed, hard-water, temperate-zone streams), as many as 1,400 nymphs have been found in one square foot of surface, and one gravel riffle has yielded as many as 33 species.

Life Cycle

The life cycle of mayflies consists of four stages: egg, nymph, subimago, and imago. Eggs, which vary widely in size and surface detail, may be oblong, oval, or rounded. Depending on the species, a female may produce fewer than 50 or more than 10,000 eggs. Eggs are laid in water and either settle to the bottom or adhere to some submerged object. They often hatch in about two weeks but may, under certain circumstances, undergo a period of varying duration in which no growth occurs. This cessation of growth, known as diapause, is a highly effective adaptation that enables the insects to avoid environmental conditions hostile to developing nymphs or to emerging winged stages.

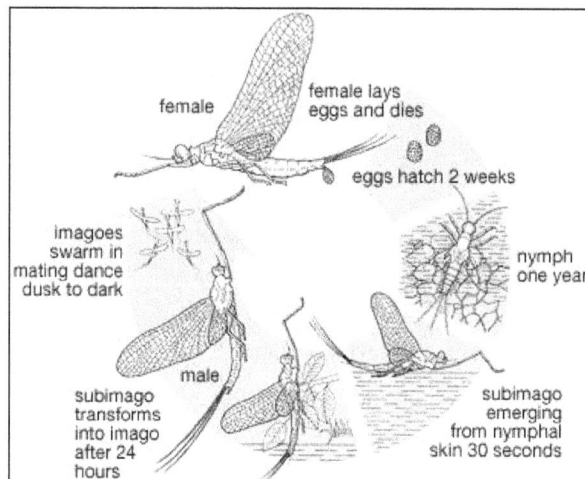

Life cycle of the mayfly (order Ephemeroptera).

Nymphal life may be as short as two weeks or as long as two years, although an annual cycle is most common. As many as 50 molts (periodic shedding of skin) may occur, depending on the species and the environment. When growth is complete, the nymphal skin splits down the back and a winged form, called the subimago, or dun, emerges. The subimago flies from the surface of the water to some sheltered resting place nearby. After an interval lasting a few minutes to several days, but usually overnight, the skin is shed for the last time, and the imago, or adult stage (sometimes called a spinner), emerges. Mayflies are the only insects that molt after developing functional wings. The subimago resembles the imago in overall appearance, although it is softer and duller in colour than the adult. The wings of the subimago, generally rather opaque, are tinted with gray, blue, yellow, or olive. Heavy pigmentation along the veins may give the wings of the subimago a mottled appearance that rarely persists in the imago. Legs and tails of the subimago are shorter than are those of the imago. It is often incorrectly assumed that the two stages are different species.

Mating and Egg Deposition

Mating takes place soon after the final molt. In most species death ensues shortly after mating and oviposition (egg deposition). Winged existence may last only a few hours, although Hexagenia males may live long enough to engage in mating flights on two successive days, and female imagos that retain their eggs may live long enough to mate on either of two successive days. Groups of male imagoes perform a mating flight, or dance, over water as dusk approaches, flying into any breeze or air current. Individuals may fly up and forward, then float downward and repeat the performance. Females soon join the swarm, rising and falling as the dance continues. The male approaches the female from below and behind and grasps her thorax with his elongated front legs. Mating is completed on the wing. After her release by the male, the female deposits her eggs and dies. A few species are ovoviviparous—i.e., eggs hatch within the body of the female generally as she floats, dying, on the surface of a stream or pond.

Methods of oviposition vary. Some species drop the rounded egg mass from a height of several feet in a manoeuvre suggestive of dive-bombing, whereas in others, the female flies low over the water's surface, striking it at intervals with the tip of her abdomen and washing off a few eggs each time she strikes the water. Still other females extrude the eggs from two oviducts as two long packets, which usually adhere to each other. They may be dropped from a foot or more above the water, but more often, the female falls to the surface with wings extended and squeezes out the eggs as she dies. In a fourth type of oviposition, the female alights on some object protruding from the water and crawls under the surface, depositing the eggs while submerged. Females, unless they drop the eggs from a height of several feet, are vulnerable to feeding fishes. Mayflies sometimes mistake blacktopped roads for streams, forming swarms over them, and drop eggs on road surfaces.

Ecology

Mayfly nymphs are preyed upon by carnivorous invertebrates and fishes. Winged stages are devoured in flight by birds, bats, and predatory insects, including dragonflies, robber flies, and hornets. When at rest, mayflies may be preyed upon by spiders, beetles, birds, and certain mammals, especially flying squirrels in North America. During their transformation to the adult stage and especially during oviposition by females, mayflies are vulnerable to predation by fishes; artificial lures used by fishermen are patterned after them.

Form and Function

Adaptations of form and function presumably determine distribution. The legs and jaws of some nymphs are modified for burrowing in silt or sand, whereas in other species, these are flattened to facilitate entering narrow crevices or clinging to bottom materials in swift currents. Long, slender legs and body adapt others for clambering on submerged vegetation. Strong swimmers are long and slender and occupy a variety of habitats. Gills may be platelike, feathery, or filamentous and may be modified for specialized functions.

Paleontology and Classification

Recognizable mayflies occur in the fossil record of the Pennsylvanian Subperiod (about 323.2 million to 298.9 million years ago), and they appear to have been abundant during the Permian (298.9

million to 252.2 million years ago). Represented largely by wing impressions, the fossil record is so incomplete that most systems of classification and interpretations of relationships are based on characteristics of recent forms, chiefly their morphology.

Distinguishing Taxonomic Features

Characteristics of the male genitalia are the most reliable means for identification of adult species. Many other features, including patterns of veins in the wings, affect generic and other higher categories of classification.

Annotated classification

Order Ephemeroptera (Mayflies)

Soft-bodied insects; life cycle consisting of 4 stages—egg, nymph, subimago, imago; wings membranous, at rest held vertically upward; hind wings reduced; mouthparts and digestive system of adults nonfunctional; only insect to molt after developing functional wings; antennae bristlelike; 3 suborders—Carapacea (armoured mayflies); Furcatergalia (forked-gill mayflies); and Pisciforma (brush-legged, flat-headed, and minnow mayflies); about 2,500 species on all continents except Antarctica.

Critical Appraisal

Various classification schemes have been proposed for Ephemeroptera. Increasingly, these schemes are based on phylogenetic relationships. However, some relationships remain to be resolved, particularly at higher levels (e.g., superfamily and family).

Dragonfly

A dragonfly is an insect belonging to the order Odonata, infraorder Anisoptera. Adult dragonflies are characterized by large, multifaceted eyes, two pairs of strong, transparent wings, sometimes with coloured patches, and an elongated body. Dragonflies can be mistaken for the related group, damselflies (Zygoptera), which are similar in structure, though usually lighter in build; however, the wings of most dragonflies are held flat and away from the body, while damselflies hold the wings folded at rest, along or above the abdomen. Dragonflies are agile fliers, while damselflies have a weaker, fluttery flight. Many dragonflies have brilliant iridescent or metallic colours produced by structural coloration, making them conspicuous in flight. An adult dragonfly's compound eyes have nearly 24,000 ommatidia each.

Fossils of very large dragonfly ancestors in the Protodonata are found from 325 million years ago (Mya) in Upper Carboniferous rocks; these had wingspans up to about 750 mm (30 in). There are about 3,000 extant species. Most are tropical, with fewer species in temperate regions.

Dragonflies are predators, both in their aquatic larval stage, when they are known as nymphs or naiads, and as adults. Several years of their lives are spent as nymphs living in fresh water; the adults may be on the wing for just a few days or weeks. They are fast, agile fliers, sometimes migrating across oceans, and often live near water. They have a uniquely complex mode of reproduction involving indirect insemination, delayed fertilization, and sperm competition. During mating, the male grasps the female at the back of the head, and the female curls her abdomen under her body to pick up sperm from the male's secondary genitalia at the front of his abdomen, forming the "heart" or "wheel" posture.

Loss of wetland habitat threatens dragonfly populations around the world. Dragonflies are represented in human culture on artifacts such as pottery, rock paintings, and Art Nouveau jewelry. They are used in traditional medicine in Japan and China, and caught for food in Indonesia. They are symbols of courage, strength, and happiness in Japan, but seen as sinister in European folklore. Their bright colours and agile flight are admired in the poetry of Lord Tennyson and the prose of H. E. Bates.

Phylogeny

Dragonflies and their relatives are an ancient group. The oldest fossils are of the Protodonata from the 325 Mya Upper Carboniferous of Europe, a group that included the largest insect that ever lived, Meganeuropsis permiana from the Early Permian, with a wingspan around 750 mm (30 in); their fossil record ends with the Permian–Triassic extinction event (about 247 Mya). The Protanisoptera, another ancestral group which lacks certain wing vein characters found in modern Odonata, lived from the Early to Late Permian age until the end Permian event, and are known from fossil wings from current day United States, Russia, and Australia, suggesting they might have been cosmopolitan in distribution. The forerunners of modern Odonata are included in a clade called the Panodonata, which include the basal Zygoptera (damselflies) and the Anisoptera (true dragonflies). Today there are some 3000 species extant around the world.

The relationships of anisopteran families are not fully resolved as of 2013, but all the families are monophyletic except the Corduliidae; the Gomphidae are a sister taxon to all other Anisoptera, the Austropetaliidae are sister to the Aeshnoidea, and the Chlorogomphidae are sister to a clade that includes the Synthemistidae and Libellulidae. On the cladogram, dashed lines indicate unresolved relationships; English names are given (in parentheses).

The giant Upper Carboniferous dragonfly ancestor, Meganeura monyi,
attained a wingspan of about 680 mm (27 in). Museum of Toulouse

Mesurupetala, Late Jurassic (Tithonian),
Solnhofen limestone, Germany

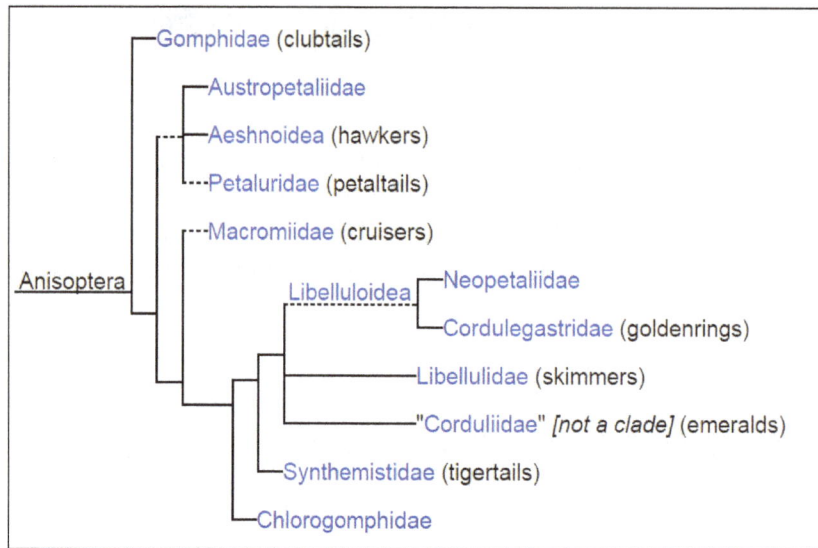

Distribution and Diversity

About 3012 species of dragonflies were known in 2010; these are classified into 348 genera in 11 families. The distribution of diversity within the bio-geographical regions are summarized below (the world numbers are not ordinary totals, as overlaps in species occur).

Family	Oriental	Neotropical	Australasian	Afrotropical	Palaearctic	Nearctic	Pacific	World
Aeshnidae	149	129	78	44	58	40	13	456
Austropetali-idae		7	4					11
Petaluridae		1	6		1	2		10
Gomphidae	364	277	42	152	127	101		980
Chlorogomphi-dae	46				5			47
Cordulegast-ridae	23	1			18			46
Neopetaliidae		1						1
Corduliidae	23	20	33	6	18	51	12	154
Libellulidae	192	354	184	251	120	105	31	1037
Macromiidae	50	2	17	37	7	10		125
Synthemistidae			37				9	46
Incertae sedis	37	24	21	15	2			99

Dragonflies live on every continent except Antarctica. In contrast to the damselflies (Zygoptera), which tend to have restricted distributions, some genera and species are spread across continents. For example, the blue-eyed darner Rhionaeschna multicolor lives all across North America, and in Central America; emperors Anax live throughout the Americas from as far north as Newfoundland to as far south as Bahia Blanca in Argentina, across Europe to central Asia, North Africa, and the Middle East. The globe skimmer Pantala flavescens is probably the most widespread dragonfly species in the world; it is cosmopolitan, occurring on all continents in the warmer regions. Most Anisoptera species are tropical, with far fewer species in temperate regions.

Some dragonflies, including libellulids and aeshnids, live in desert pools, for example in the Mojave Desert, where they are active in shade temperatures between 18 and 45 °C (64.4 to 113 °F); these insects were able to survive body temperatures above the thermal death point of insects of the same species in cooler places.

Dragonflies live from sea level up to the mountains, decreasing in species diversity with altitude. Their altitudinal limit is about 3700 m, represented by a species of Aeshna in the Pamirs.

Dragonflies become scarce at higher latitudes. They are not native to Iceland, but individuals are occasionally swept in by strong winds, including a Hemianax ephippiger native to North Africa, and an unidentified darter species. In Kamchatka, only a few species of dragonfly including the treeline emerald Somatochlora arctica and some aeshnids such as Aeshna subarctica are found, possibly because of the low temperature of the lakes there. The treeline emerald also lives in northern Alaska, within the Arctic Circle, making it the most northerly of all dragonflies.

An aggregation of globe skimmers, Pantala
flavescens, during migration

Description

Dragonflies (suborder Anisoptera) are heavy-bodied, strong-flying insects that hold their wings horizontally both in flight and at rest. By contrast, damselflies (suborder Zygoptera) have slender bodies and fly more weakly; most species fold their wings over the abdomen when stationary, and the eyes are well separated on the sides of the head.

An adult dragonfly has three distinct segments, the head, thorax, and abdomen as in all insects. It has a chitinous exoskeleton of hard plates held together with flexible membranes. The head is large with very short antennae. It is dominated by the two compound eyes, which cover most of its surface. The compound eyes are made up of ommatidia, the numbers being greater in the larger species. Aeshna interrupta has 22650 ommatidia of two varying sizes, 4500 being large. The facets facing downward tend to be smaller. Petalura gigantea has 23890 ommatidia of just one size. These facets provide complete vision in the frontal hemisphere of the dragonfly. The compound eyes meet at the top of the head (except in the Petaluridae and Gomphidae, as also in the genus Epiophlebia). Also, they have three simple eyes or ocelli. The mouthparts are adapted for biting with a toothed jaw; the flap-like labrum, at the front of the mouth, can be shot rapidly forward to catch prey. The head has a system for locking it in place that consists of muscles and small hairs on the back of the head that grip structures on the front of the first thoracic segment. This arrester system is unique to the Odonata, and is activated when feeding and during tandem flight.

Damselflies, like this Ischnura senegalensis, are slenderer in build than dragonflies, and most hold their wings closed over their bodies.

The thorax consists of three segments as in all insects. The prothorax is small and is flattened dorsally into a shield-like disc which has two transverse ridges. The mesothorax and metathorax are fused into a rigid, box-like structure with internal bracing, and provides a robust attachment for the powerful wing muscles inside it. The thorax bears two pairs of wings and three pairs of legs. The wings are long, veined, and membranous, narrower at the tip and wider at the base. The hindwings are broader than the forewings and the venation is different at the base. The veins carry haemolymph, which is analogous to blood in vertebrates and carries out many similar functions, but which also serves a hydraulic function to expand the body between nymphal stages (instars) and to expand and stiffen the wings after the adult emerges from the final nymphal stage. The leading edge of each wing has a node where other veins join the marginal vein, and the wing is able to flex at this point. In most large species of dragonflies, the wings of females are shorter and broader than those of males. The legs are rarely used for walking, but are used to catch and hold prey, for perching, and for climbing on plants. Each has two short basal joints, two long joints, and a three-jointed foot, armed with a pair of claws. The long leg joints bear rows of spines, and in males, one row of spines on each front leg is modified to form an "eyebrush", for cleaning the surface of the compound eye.

Head of the blue Aeshne dragonfly

The abdomen is long and slender and consists of 10 segments. There are three terminal appendages on segment 10; a pair of superiors (claspers) and an inferior. The second and third segments are enlarged, and in males, on the underside of the second segment has a cleft, forming the secondary genitalia consist of lamina, hamule, genital lobe and penis. There are remarkable variations in the presence and the form of the penis and the related structures, the flagellum, cornua and genital lobes. Sperm is produced at the 9th segment and is transferred to the secondary genitalia prior to mating. The male holds the female behind the head using a pair of claspers on the terminal segment. In females, the genital opening is on the underside of the eighth segment and is covered by a simple flap (vulvar lamina) or an ovipositor, depending on species and the method of egg-laying. Dragonflies having simple flap shed the eggs in water, mostly in flight. Dragonflies having ovipositor, use it to puncture soft tissues of plants and place the eggs singly in each puncture they made.

Migrant hawker, Aeshna mixta, has the long
slender abdomen of aeshnid dragonflies.

Dragonfly nymphs vary in form with species and are loosely classed into claspers, sprawlers, hiders, and burrowers. The first instar is known as a prolarva, a relatively inactive stage from which it quickly moults into the more active nymphal form. The general body plan is similar to that of an adult, but the nymph lacks wings and reproductive organs. The lower jaw has a huge, extensible labium, armed with hooks and spines, which is used for catching prey. This labium is folded under the body at rest and struck out at great speed by hydraulic pressure created by the abdominal muscles. Whereas damselfly nymphs have three feathery external gills, dragonfly nymphs have internal gills, located around the fourth and fifth abdominal segments. Water is pumped in and out of the abdomen through an opening at the tip. The naiads of some clubtails (Gomphidae) that burrow into the sediment, have a snorkel-like tube at the end of the abdomen enabling them to draw in clean water while they are buried in mud. Naiads can forcefully expel a jet of water to propel themselves with great rapidity.

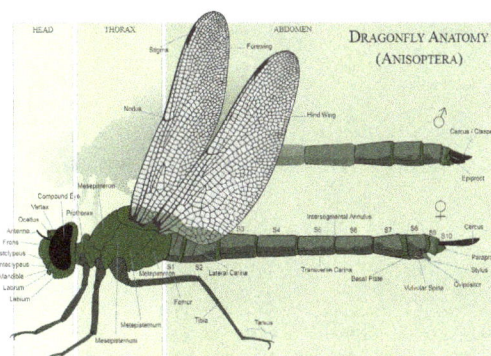

Anatomy of a dragonfly

Coloration

Many adult dragonflies have brilliant iridescent or metallic colours produced by structural coloration, making them conspicuous in flight. Their overall coloration is often a combination of yellow, red, brown, and black pigments, with structural colours. Blues are typically created by microstructures in the cuticle that reflect blue light. Greens often combine a structural blue with a yellow pigment. Freshly emerged adults, known as tenerals, are often pale-coloured and obtain their typical colours after a few days, some have their bodies covered with a pale blue, waxy powderiness called pruinosity; it wears off when scraped during mating, leaving darker areas.

Iridescent structural coloration in a dragonfly's eyes

Some dragonflies, such as the green darner, Anax junius, have a noniridescent blue which is produced structurally by scatter from arrays of tiny spheres in the endoplasmic reticulum of epidermal cells underneath the cuticle.

The wings of dragonflies are generally clear, apart from the dark veins and pterostigmata. In the chasers (Libellulidae), however, many genera have areas of colour on the wings: for example, groundlings (Brachythemis) have brown bands on all four wings, while some scarlets (Crocothemis) and dropwings (Trithemis) have bright orange patches at the wing bases. Some aeshnids such as the brown hawker (Aeshna grandis) have translucent, pale yellow wings.

Dragonfly nymphs are usually a well-camouflaged blend of dull brown, green, and grey.

Male green darner, Anax junius has non-iridescent structural blue; the female (below) lacks the colour.

Ecology

Dragonflies and damselflies are predatory both in the aquatic nymphal and adult stages. Nymphs feed on a range of freshwater invertebrates and larger ones can prey on tadpoles and small fish. Adults capture insect prey in the air, making use of their acute vision and highly controlled flight. The mating system of dragonflies is complex and they are among the few insect groups that have a system of indirect sperm transfer along with sperm storage, delayed fertilization, and sperm competition.

Adult males vigorously defend territories near water; these areas provide suitable habitat for the larvae to develop, and for females to lay their eggs. Swarms of feeding adults aggregate to prey on swarming prey such as emerging flying ants or termites.

Dragonflies as a group occupy a considerable variety of habitats, but many species, and some families, have their own specific environmental requirements. Some species prefer flowing waters, while others prefer standing water. For example, the Gomphidae (clubtails) live in running water, and the Libellulidae (skimmers) live in still water. Some species live in temporary water pools and are capable of tolerating changes in water level, desiccation, and the resulting variations in temperature, but some genera such as Sympetrum (darters) have eggs and larvae that can resist drought and are stimulated to grow rapidly in warm, shallow pools, also often benefiting from the absence of predators there. Vegetation and its characteristics including submerged, floating, emergent, or waterside are also important. Adults may require emergent or waterside plants to use as perches; others may need specific submerged or floating plants on which to lay eggs. Requirements may be highly specific, as in Aeshna viridis (green hawker), which lives in swamps with the water-soldier, Stratiotes aloides. The chemistry of the water, including its trophic status (degree of enrichment with nutrients) and pH can also affect its use by dragonflies. Most species need moderate conditions, not too eutrophic, not too acid; a few species such as Sympetrum danae (black darter) and Libellula quadrimaculata (four-spotted chaser) prefer acidic waters such as peat bogs, while others such as Libellula fulva (scarce chaser) need slow-moving, eutrophic waters with reeds or similar waterside plants.

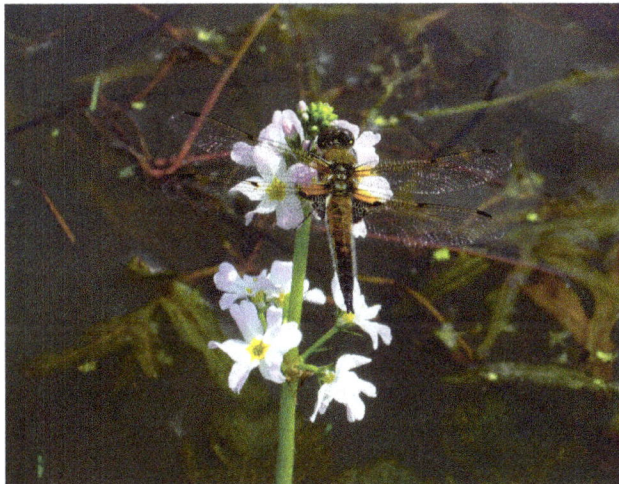

Habitat preference: a four-spotted chaser, Libellula quadrimaculata
on an emergent plant, the water violet Hottonia palustris,
with submerged vegetation in the background

Behaviour

Many dragonflies, particularly males, are territorial. Some defend a territory against others of their own species, some against other species of dragonfly and a few against insects in unrelated groups. A particular perch may give a dragonfly a good view over an insect-rich feeding ground, and the blue dasher (Pachydiplax longipennis) jostles other dragonflies to maintain the right to alight there.

Defending a breeding territory is fairly common among male dragonflies, especially among species that congregate around ponds in large numbers. The territory contains desirable features such as a sunlit stretch of shallow water, a special plant species, or a particular substrate necessary for egg-laying. The territory may be small or large, depending on its quality, the time of day, and the number of competitors, and may be held for a few minutes or several hours. Some dragonflies signal ownership with striking colours on the face, abdomen, legs, or wings. The common whitetail (Plathemis lydia) dashes towards an intruder holding its white abdomen aloft like a flag. Other dragonflies engage in aerial dogfights or high-speed chases. A female must mate with the territory holder before laying her eggs. There is also conflict between the males and females. Females may sometimes be harassed by males to the extent that it affects their normal activities including foraging and in some dimorphic species females have evolved multiple forms with some forms appearing deceptively like males. In some species females have evolved behavioural responses such as feigning death to escape the attention of males.

Reproduction

Mating in dragonflies is a complex, precisely choreographed process. First, the male has to attract a female to his territory, continually driving off rival males. When he is ready to mate, he transfers a packet of sperm from his primary genital opening on segment 9, near the end of his abdomen, to his secondary genitalia on segments 2–3, near the base of his abdomen. The male then grasps the female by the head with the claspers at the end of his abdomen; the structure of the claspers varies between species, and may help to prevent interspecific mating. The pair flies in tandem with the male in front, typically perching on a twig or plant stem. The female then curls her abdomen downwards and forwards under her body to pick up the sperm from the male's secondary genitalia, while the male uses his "tail" claspers to grip the female behind the head: this distinctive posture is called the "heart" or "wheel"; the pair may also be described as being "in cop".

Egg-laying (ovipositing) involves not only the female darting over floating or waterside vegetation to deposit eggs on a suitable substrate, but also the male hovering above her or continuing to clasp her and flying in tandem. The male attempts to prevent rivals from removing his sperm and inserting their own, something made possible by delayed fertilisation and driven by sexual selection. If successful, a rival male uses his penis to compress or scrape out the sperm inserted previously; this activity takes up much of the time that a copulating pair remains in the heart posture. Flying in tandem has the advantage that less effort is needed by the female for flight and more can be expended on egg-laying, and when the female submerges to deposit eggs, the male may help to pull her out of the water.

Egg-laying takes two different forms depending on the species. The female in some families has a sharp-edged ovipositor with which she slits open a stem or leaf of a plant on or near the water, so she can push her eggs inside. In other families such as clubtails (Gomphidae), cruisers (Macromii-dae), emeralds (Corduliidae), and skimmers (Libellulidae), the female lays eggs by tapping the surface of the water repeatedly with her abdomen, by shaking the eggs out of her abdomen as she flies along, or by placing the eggs on vegetation. In a few species, the eggs are laid on emergent plants above the water, and development is delayed until these have withered and become immersed.

Mating pair of marsh skimmers,
Orthetrum luzonicum, forming a "heart"

Life Cycle

Dragonflies are hemimetabolous insects; they do not have a pupal stage and undergo an incomplete metamorphosis with a series of nymphal stages from which the adult emerges. Eggs laid inside plant tissues are usually shaped like grains of rice, while other eggs are the size of a pinhead, ellipsoidal, or nearly spherical. A clutch may have as many as 1500 eggs, and they take about a week to hatch into aquatic nymphs or naiads which moult between six and 15 times (depending on species) as they grow. Most of a dragonfly's life is spent as a nymph, beneath the water's surface. The nymph extends its hinged labium (a toothed mouthpart similar to a lower mandible, which is sometimes termed as a "mask" as it is normally folded and held before the face) that can extend forward and retract rapidly to capture prey such as mosquito larvae, tadpoles, and small fish. They breathe through gills in their rectum, and can rapidly propel themselves by suddenly expelling water through the anus. Some naiads, such as the later stages of Antipodophlebia asthenes, hunt on land.

Nymph of emperor dragonfly, Anax imperator

The larval stage of dragonflies lasts up to five years in large species, and between two months and three years in smaller species. When the naiad is ready to metamorphose into an adult, it stops feeding and makes its way to the surface, generally at night. It remains stationary with its head out of the water, while its respiration system adapts to breathing air, then climbs up a reed or other emergent plant, and moults (ecdysis). Anchoring itself firmly in a vertical position with its claws, its skin begins to split at a weak spot behind the head. The adult dragonfly crawls out of its larval skin, the exuvia, arching backwards when all but the tip of its abdomen is free, to allow its exoskeleton to harden. Curling back upwards, it completes its emergence, swallowing air, which plumps out its body, and pumping haemolymph into its wings, which causes them to expand to their full extent.

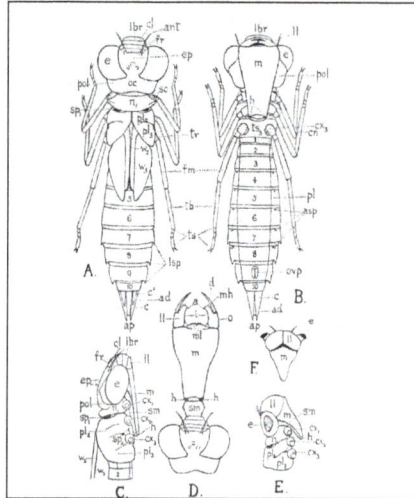

Parts of a dragonfly nymph including the labial "mask"

Dragonflies in temperate areas can be categorized into two groups, an early group and a later one. In any one area, individuals of a particular "spring species" emerge within a few days of each other. The springtime darner (Basiaeschna janata), for example, is suddenly very common in the spring, but disappears a few weeks later and is not seen again until the following year. By contrast, a "summer species" emerges over a period of weeks or months, later in the year. They may be seen on the wing for several months, but this may represent a whole series of individuals, with new adults hatching out as earlier ones complete their short lifespans which is an average of 7 months.

Ecdysis: Emperor dragonfly, Anax imperator, newly emerged
and still soft, holding on to its dry exuvia, and expanding its wings

Sex Ratios

The sex ratio of male to female dragonflies varies both temporally and spatially. Adult dragonflies have a high male-biased ratio at breeding habitats. The male-bias ratio has contributed partially to the females using different habitats to avoid male harassment. As seen in Hine's emerald dragonfly (Somatochlora hineana), male populations use wetland habitats, while females use dry meadows and marginal breeding habitats, only migrating to the wetlands to lay their eggs or to find mating partners. Unwanted mating is energetically costly for females because it affects the amount of time that they are able to spend foraging.

Flight

Dragonflies are powerful and agile fliers, capable of migrating across the sea, moving in any direction, and changing direction suddenly. In flight, the adult dragonfly can propel itself in six directions: upward, downward, forward, backward, to left and to right. They have four different styles of flight: A number of flying modes are used that include counter-stroking, with forewings beating 180° out of phase with the hindwings, is used for hovering and slow flight. This style is efficient and generates a large amount of lift; phased-stroking, with the hindwings beating 90° ahead of the forewings, is used for fast flight. This style creates more thrust, but less lift than counter-stroking; synchronised-stroking, with forewings and hindwings beating together, is used when changing direction rapidly, as it maximises thrust; and gliding, with the wings held out, is used in three situations: free gliding, for a few seconds in between bursts of powered flight; gliding in the updraft at the crest of a hill, effectively hovering by falling at the same speed as the updraft; and in certain dragonflies such as darters, when "in cop" with a male, the female sometimes simply glides while the male pulls the pair along by beating his wings.

The wings are powered directly, unlike most families of insects, with the flight muscles attached to the wing bases. Dragonflies have a high power/weight ratio, and have been documented accelerating at 4 G linearly and 9 G in sharp turns while pursuing prey.

Brown hawker, Aeshna grandis in flight: The hindwings are about 90° out of phase with the forewings at this instant, suggesting fast flight.

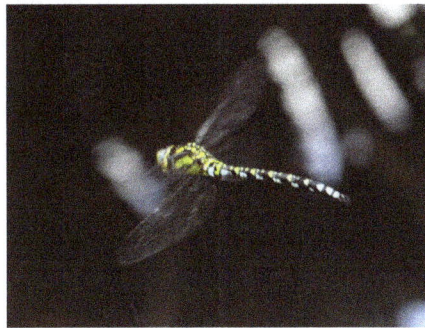

Southern hawker, Aeshna cyanea: its wings at this instant are synchronised for agile flight.

Dragonflies generate lift in at least four ways at different times, including classical lift like an aircraft wing; supercritical lift with the wing above the critical angle, generating high lift and using very short strokes to avoid stalling; and creating and shedding vortices. Some families appear to use special mechanisms, as for example the Libellulidae which take off rapidly, their wings beginning pointed far forward and twisted almost vertically. Dragonfly wings behave highly dynamically during flight, flexing and twisting during each beat. Among the variables are wing curvature, length and speed of stroke, angle of attack, forward/back position of wing, and phase relative to the other wings.

Red-veined darters (Sympetrum fonscolombii)
flying "in cop" (male ahead)

Flight Speed

Old and unreliable claims are made that dragonflies such as the southern giant darner can fly up to 97 km/h (60 mph). However, the greatest reliable flight speed records are for other types of insects. In general, large dragonflies like the hawkers have a maximum speed of 36–54 km/h (22–34 mph) with average cruising speed of about 16 km/h (9.9 mph). Dragonflies can travel at 100 body-lengths per second in forward flight, and three lengths per second backwards.

Motion Camouflage

In high-speed territorial battles between male Australian emperors (Hemianax papuensis), the fighting dragonflies adjust their flight paths to appear stationary to their rivals, minimizing the chance of being detected as they approach. To achieve the effect, the attacking dragonfly flies towards his rival, choosing his path to remain on a line between the rival and the start of his attack path. The attacker thus looms larger as he closes on the rival, but does not otherwise appear to move. Researchers found that six of 15 encounters involved motion camouflage.

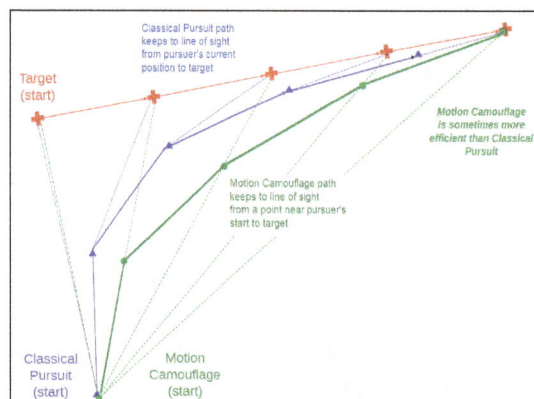

The principle of motion camouflage

Temperature Control

The flight muscles need to be kept at a suitable temperature for the dragonfly to be able to fly. Being cold-blooded, they can raise their temperature by basking in the sun. Early in the morning, they may choose to perch in a vertical position with the wings outstretched, while in the middle of the day, a horizontal stance may be chosen. Another method of warming up used by some larger dragonflies is wing-whirring, a rapid vibration of the wings that causes heat to be generated in the flight muscles. The green darner (Anax junius) is known for its long-distance migrations, and often resorts to wing-whirring before dawn to enable it to make an early start.

Becoming too hot is another hazard, and a sunny or shady position for perching can be selected according to the ambient temperature. Some species have dark patches on the wings which can provide shade for the body, and a few use the obelisk posture to avoid overheating. This behaviour involves doing a "handstand", perching with the body raised and the abdomen pointing towards the sun, thus minimising the amount of solar radiation received. On a hot day, dragonflies sometimes adjust their body temperature by skimming over a water surface and briefly touching it, often three times in quick succession. This may also help to avoid desiccation.

Feeding

Adult dragonflies hunt on the wing using their exceptionally acute eyesight and strong, agile flight. They are almost exclusively carnivorous, eating a wide variety of insects ranging from small midges and mosquitoes to butterflies, moths, damselflies, and smaller dragonflies. A large prey item is subdued by being bitten on the head and is carried by the legs to a perch. Here, the wings are discarded and the prey usually ingested head first. A dragonfly may consume as much as a fifth of its body weight in prey per day. Dragonflies are also some of the insect world's most efficient hunters, catching up to 95% of the prey they pursue.

The larvae are voracious predators, eating most living things that are smaller than they are. Their staple diet is mostly bloodworms and other insect larvae, but they also feed on tadpoles and small fish. A few species, especially those that live in temporary waters, are likely to leave water. Nymphs of Cordulegaster bidentata sometimes hunt small arthropods on the ground at night.

Common clubtail, Gomphus vulgatissimus, with prey

Predators and Parasites

Although dragonflies are swift and agile fliers, some predators are fast enough to catch them. These include falcons such as the American kestrel, the merlin, and the hobby; nighthawks, swifts, flycatchers and swallows also take some adults; some species of wasps, too, prey on dragonflies, using them to provision their nests, laying an egg on each captured insect. In the water, various species of ducks and herons eat dragonfly larvae and they are also preyed on by newts, frogs, fish, and water spiders. Amur falcons, which migrate over the Indian Ocean at a period that coincides with the migration of the globe skimmer dragonfly, Pantala flavescens, may actually be feeding on them while on the wing.

Dragonflies are affected by three major groups of parasites: water mites, gregarine protozoa, and trematode flatworms (flukes). Water mites, Hydracarina, can kill smaller dragonfly larvae, and may also be seen on adults. Gregarines infect the gut and may cause blockage and secondary infection. Trematodes are parasites of vertebrates such as frogs, with complex life cycles often involving a period as a stage called a cercaria in a secondary host, a snail. Dragonfly nymphs may swallow cercariae, or these may tunnel through a nymph's body wall; they then enter the gut and form a cyst or metacercaria, which remains in the nymph for the whole of its development. If the nymph is eaten by a frog, the amphibian becomes infected by the adult or fluke stage of the trematode.

Southern red-billed hornbill with a captured dragonfly in its bill

Conservation

Most odonatologists live in temperate areas and the dragonflies of North America and Europe have been the subject of much research. However, the majority of species live in tropical areas and have been little studied. With the destruction of rainforest habitats, many of these species are in danger of becoming extinct before they have even been named. The greatest cause of decline is forest clearance with the consequent drying up of streams and pools which become clogged with silt. The damming of rivers for hydroelectric schemes and the drainage of low-lying land has reduced suitable habitat, as has pollution and the introduction of alien species.

In 1997, the International Union for Conservation of Nature set up a status survey and conservation action plan for dragonflies. This proposes the establishment of protected areas around the world and the management of these areas to provide suitable habitat for dragonflies. Outside these areas, encouragement should be given to modify forestry, agricultural, and industrial practices to enhance conservation. At the same time, more research into dragonflies needs to be done, consideration should be given to pollution control and the public should be educated about the importance of biodiversity.

Habitat degradation has reduced dragonfly populations across the world, for example in Japan. Over 60% of Japan's wetlands were lost in the 20th century, so its dragonflies now depend largely on rice fields, ponds, and creeks. Dragonflies feed on pest insects in rice, acting as a natural pest control. Dragonflies are steadily declining in Africa, and represent a conservation priority.

The dragonfly's long lifespan and low population density makes it vulnerable to disturbance, such as from collisions with vehicles on roads built near wetlands. Species that fly low and slow may be most at risk.

Dragonflies are attracted to shiny surfaces that produce polarization which they can mistake for water, and they have been known to aggregate close to polished gravestones, solar panels, automobiles, and other such structures on which they attempt to lay eggs. These can have a local impact on dragonfly populations; methods of reducing the attractiveness of structures such as solar panels are under experimentation.

Damselfly

Damselfly is the common name for any of the predaceous insects comprising the suborder Zygoptera of the order Odonata, characterized by an elongated body, large multifaceted eyes that are widely separated, and two pairs of strong transparent wings, which at rest typically are held folded together above the abdomen or held slightly open above the abdomen. Damselflies are similar to dragonflies (suborder or infraorder Anisoptera) but differ in several ways, including the fact that dragonflies at rest typically hold their wings out to the side or out and downward.

Damselflies provide important ecological and aesthetic values. Widely distributed, found on every continent except Antarctica, both the larvae and adults are key components of aquatic and terrestrial food chains, serving as both predator and prey in both systems, and helping to control insects pests, such as mosquitoes. For humans, they are a popular subject of art and culture in various nations, notably Japan, and their grace, often striking colors, and unique mating behaviors add to the beauty of nature.

Overview and Description

As with all members of the arthropod class Insecta, damselflies have three pairs of jointed appendages, exposed mouth parts, an exoskeleton, a segmented abdomen that lacks any legs or wings, and one pair of antennae on the head.

Damselflies and dragonflies comprise the order Odonata, a taxon of about 6,500 insects placed into just over 600 genera. Members of Odonata are characterized by large, compound eyes, chewing mouth parts, a long and slender abdomen, and multi-veined wings that are typically clear or transparent. They also have large rounded heads, legs that facilitate catching prey (other insects) in flight, two pairs of long, transparent wings that move independently, and elongated, ten-segmented abdomens. In most families of Odonata, the wings, which are large, multi-veined, and slender, have an opaque structure on the leading edge near the tip of the wing, called the pterostigma. Males have unique secondary genitalia on the underside of the second and third abdominal segments, which are distinct from the actual genital opening located near the tip of the abdomen.

Within Odonata, the damselflies are placed in suborder Zygoptera, while the dragonflies are placed in suborder Aniosptera, or in suborder Epiprocta with true dragonflies in infraorder Anisoptera.

Although generally fairly similar, with an elongated body, large multifaceted eyes, and two pairs of strong transparent wings, adults of damselflies and dragonflies differ in several, easily recognizable traits. The large, compound eyes of damselflies typically have a gap between them. In dragonflies, the eyes typically occupy much of the animal's head, touching (or nearly touching) each other across the face (with notable exceptions to this being in the Petaluridae (petaltails) and the Gomphidae (clubtails) families). Damselflies also tend to be less robust than dragonflies, even appearing rather weak in flight, and when at rest hold their wings either folded together back above the abdomen or held slightly open above (such as in the family Lestidae). Dragonflies have fairly robust bodies, are strong fliers, and at rest hold their wings either out to the side or out and downward (or even somewhat forward). There are exceptions to this, as some zygopteran families have wings that are held horizontally at rest, and in one anispteran genus the wings are held vertically together above the abdomen, like damselflies. Another difference between damselfies and dragonflies relates to the hind wing. The hind wing of the damselfly is essentially similar to the fore wing, while the hind wing of the dragonfly broadens near the base, caudal to the connecting point at the body.

In damselflies, each of the two pairs of wings are nearly exactly equal in size, shape, and venation; there may be very numerous crossveins or rather few. Damselfly wingspans range from just 20 millimeters (0.8 inches) in Agriocnemis femina up to 190 centimeters (7.5 inches) in the giant Central American damselfly, Megaloprepus coerulatus. Some very large fossil species have been discovered as well.

The larval stage of damselflies (and dragonflies) are characterized by a conspicuous grasping labium used for catching prey. This lower lip "mask" is held at rest in a folded position underneath the head and thorax, with one end extending back as far or further than the front legs, with the anterior portion sometimes forward enough in some species to cover the lower part of the face, below the compound eyes. In capturing prey, the labium is extended rapidly forward to grasp the prey with paired palps, like grasping hands.

Head of a damselfly

Damselfly, Ischnura senegalensis

Life Cycle

As a member of Odonata, the damselfly life cycle is similar to that of the dragonfly.

Damselflies undergo incomplete metamorphosis. Incomplete metamorphosis, also called hemimetabolism, simple metamorphosis, gradual metamorphosis, or hemimetaboly, is a term applied to those processes in which the larvae resembles the adults somewhat, as they have compound eyes, developed legs, and wing stubs visible on the outside, but the juvenile forms are smaller and, if the adult has wings, lack wings. In this mode of development, there are three distinct stages: The egg, nymph, and the adult stage, or imago. These groups go through gradual changes; there is no pupal stage. In hemimetabolism, the development of larva often proceeds in repeated stages of growth and ecdysis (molting); these stages are called instars.

Mating damselflies

In damselflies, as with dragonflies, the life cycle typically has an aquatic stage. The female lays eggs in water, sometimes in underwater vegetation, or high in trees in bromeliads and other water-filled cavities. The aquatic nymphs are carnivorous, feeding on daphnia, mosquito larvae, and various other small aquatic organisms. They are non-discriminate predators, eating any animal as large as or smaller than themselves, including tadpoles and fish fry, and even members of their own species. The gills of damselfly nymphs are large and external, resembling three fins at the end of the abdomen. After molting several times, the winged adult emerges and eats flies, mosquitoes, and other small insects. Some of the larger tropical species are known to feed on spiders, hovering near the web and simply plucking the spider from its perch.

Common blue damselfly eating a leafhopper

Neoptera

Neoptera is a major taxonomic group of insects that includes almost all the winged insects and specifically those considered to be related by the ability to fold their wings back over their abdomen. Traditionally, they are one of two major groups within the subclass Pterygota (the winged insects), the other being Paleoptera, which lack the ability to flex their wings in this manner.

Some groups within Neoptera do not have the ability to fold their wings back over their abdomen, such as various butterflies and moths, but this is considered to be a feature that was lost during evolutionary history. This reflects the importance of lineage in modern classifications of organisms. With the advent of the theory of descent with modification, relatedness according to evolutionary lineage has been the primary consideration in classifying organisms. Likewise, the subclass Pterygota, which comprises the winged insects, also includes those species that do not have wings but in which it is assumed their ancestors did.

Neopterous insects include such as the beetles, flies, wasps, butterflies, true bugs, lice, bees, fleas, ants, stoneflies, grasshoppers, mantids, and cockroaches.

Insects, which are invertebrates comprising the Class Insecta, are the largest and (on land) most widely distributed taxon (taxonomic unit) within the Phylum Arthropoda. As arthropods, insects have jointed appendages, an exoskeleton (hard, external covering), segmented body, ventral nervous system, digestive system, open circulatory system, and specialized sensory receptors. Insects are distinguished from other arthropods by having three pairs of jointed legs; an abdomen that is divided into 11 segments and lacks any legs or wings; and a body separated into three parts (head, thorax, and abdomen), with one pair of antennae on the head. The true insects (that is, species classified in the Class Insecta) are also distinguished from all other arthropods in part by having ectognathous, or exposed, mouthparts.

Most species of insects, but by no means all, have wings as adults. Winged insects are placed in the Subclass Pterygota. (Wingless insects, such as the silverfishes and bristletails, are placed in the subclass Apterygota.) Pterygota also includes some insect groups that are "secondarily wingless"; that is, it is considered that the ancestors of these insects had wings but were lost through the process of descent with modification.

Neoptera are those members of Pterygota that are able to fold their wings back over their abdomen, as a result of special structures at the base of their wings. A key component of this folding mechanism is the pleural wing-folding muscle and the third axillary sclerite. Neoptera generally is considered an "infraclass." Those insects that are not able to fold their wings in this manner—such as the mayflies and the order Odonata (dragonflies, damselflies), are placed in the infraclass Paleoptera. Some insects placed in Neoptera are not able to fold their wings back but this is considered to have been a feature that their ancestors had and was lost.

Subdivisions of Infraclass Neoptera

The Neoptera may be subdivided in various ways. The Integrated Taxonomic Information System (ITIS) lumps all neopteran orders together in this infraclass without subdivision. Other authorities recognize several superorders within it.

Almost universally accepted as two major divisions of Neoptera are the Exopterygota and the Endopterygota. The Exopterygota are hemimetabolous neopterans (incomplete metamorphosis) in which the wing buds are already externally visible before the adult stage and in which no pupa or chrysalis stage occurs. The Endopterygota are holometabolous insects (complete metamorphosis, with distinctive larval, pupal, and adult stages) in which the wings develop inside the body during the larval stage and only become external appendages during the pupa or chrysalis stage. Endopterygota literally means "internal winged forms" while Exopterygota means "external winged forms," indicating whether the wing buds are evident externally in the later immature stages (in instars before the penultimate) or whether the future wing tissues are entirely internalized and make their first appearance in the penultimate (pupal) stage.

Although members of Exopterygota, such as true bugs, develop wings on the outside of their bodies without going through a true pupal stage, a few have something resembling a pupa (e.g., Aleyrodidae).

Neoptera also may be subdivided into the Endopterygota (insects with complete metamorphosis, such as beetles, flies, wasps, and butterflies), the Hemipteroid Assemblage (bugs, lice, and thrips), and the "lower Neoptera" (the many other living orders, such as Plecoptera or stoneflies, Orthoptera including grasshoppers, and Dictyoptera, including mantids and cockroaches). Another name for Endoterygota is Holometabola, indicating that these species go through complete metamorphosis.

As of recently, there are several attempts to resolve the neopteran diversity further. While this appears to be less controversial than in the (apparently paraphyletic) "Palaeoptera," there are nonetheless lots of unresolved questions. For example, the hymenopterans, traditionally considered highly advanced due to their intricate social systems, seem to be far more basal among the Endopterygota, as suggested by their relatively plesiomorphic anatomy and molecular data. The exact position of the proposed Dictyoptera is also uncertain, namely whether they are better considered Exopterygota or basal neopterans.

Termite

Termite is any of a group of cellulose-eating insects, the social system of which shows remarkable parallels with those of ants and bees, although it has evolved independently. Even though termites are not closely related to ants, they are sometimes referred to as white ants. Phylogenetic studies have shown that the closest relative to the termite is the cockroach; for this reason termites are sometimes placed in the order Dictyoptera, which also contains the mantids.

General Features

Distribution and Abundance

Termites, which number about 2,750 species, are distributed widely, reaching their greatest abundance in numbers and species in tropical rainforests around the world. In North America termites are found as far north as Vancouver, British Columbia (Zootermopsis), on the Pacific coast, and Maine and eastern Canada (Reticulitermes) on the Atlantic coast. In Europe the northern limit of natural distribution is reached by Reticulitermes lucifugus on the Atlantic coast of France, although an introduced species, Reticulitermes flavipes, occurs as far north as Hamburg, Germany. The known European species of termites have a predominantly Mediterranean distribution and do not occur naturally in Great Britain, Scandinavia, Switzerland, Germany, or northern Russia. In the Far East Reticulitermes speratus ranges as far north as South Korea, Peking, and northern Japan. Termites occur also in the Cape region of South Africa, Australia, Tasmania, and New Zealand.

In addition to naturally occurring termites, many species have been inadvertently transported by humans from their native habitats to new parts of the world. Termites, particularly Cryptotermes and Coptotermes, have been accidentally transported in wooden articles such as shipping crates, boat timbers, lumber, and furniture. Because dry-wood termites (e.g., Cryptotermes species) live in small colonies in wood and tolerate long periods of dryness, they can survive in seasoned wood and furniture and can easily be transported over long distances. Members of the family Rhinotermitidae (e.g., Coptotermes) require access to moisture and cannot survive prolonged dry periods. Coptotermes formosanus, widely distributed in Japan, Taiwan, and South China, has been introduced into Sri Lanka (Ceylon), the Pacific islands, South Africa, East Africa, Hawaii, California, and the southern United States. C. formosanus is unusual for the family in that it can survive without direct soil contact as long as a moisture source is present. In the United States the species has been found to have well-established colonies in the upper reaches of buildings, using small leaks in the roof as a moisture source. A termite native to the United States, Reticulitermes flavipes, was found in the hothouses of the Royal Palace in Schönbrunn, in Vienna, and the species was reported and described in that location before it was discovered in the United States. The termites presumably had been shipped from North America in wooden containers of decorative potted plants.

Importance

Termites are important in two ways. They are destructive when they feed upon, and often destroy, wooden structures or vegetable matter valuable to humans. Introduced species, because they are not so well equipped as native species to adapt to changes in their new environments, tend to seek shelter in protected, man-made environments such as buildings and thus are likely to become the most serious pests, causing significant damage to houses and wooden furnishings. Some termites feed on living plant materials and can become serious crop pests. Termites are also extremely beneficial in that they help to convert plant cellulose into substances that can be recycled into the ecosystem to support new growth.

Although only about 10 percent of the 2,750 known termite species have been reported as pests, many of these cause severe and extremely costly damage. For effective control, it is essential to determine whether the pest is a subterranean or a wood-dwelling species, as treatment methods differ.

Subterranean termites are dependent on contact with soil moisture and normally reach the wood in man-made structures through the foundations. The most common traditional control used around a structure is to flood a shallow trench with an insecticide and cover it with soil. Insecticides also are useful around cracks and crevices in foundations. A recent development has been to establish permanent monitoring stations around the perimeter of a foundation. When termites are discovered in a station, the wooden "bait" is replaced with a cellulose material containing a chitin inhibitor that is consumed by the foraging workers and returned to the colony and fed to others. This material disrupts normal formation of the exoskeleton (cuticle) of molting workers (the only caste capable of molting), resulting in death of the workers and, eventually, the rest of the colony. Construction and design practices that can prevent the initial entry of subterranean termites into a structure include the use of pressure-treated wood, treated concrete foundation blocks, and reinforced concrete foundations that extend at least 15.2 cm (6 inches) above the ground and have no cracks or contact with any outside wood. Removal of scrap lumber from building sites will also reduce the termite population.

Dry-wood termites nest in the wood on which they feed and do not invade a structure from the soil. Because their colonies are within the structure, they are difficult to control. Preventive measures include the use of chemically treated wood in building construction and the use of paint or other durable finish to seal cracks in wood surfaces. Fumigation is the most effective method for eliminating a dry-wood termite infestation. Another method is to place insecticide into small holes drilled into galleries of infested wood.

Colony Organization

The termite society, or colony, is a highly organized and integrated unit. There is a caste system with division of labour based on the colony members' structure, function, and behaviour. The major castes in the colony are the reproductive, soldier, and worker castes. Soldiers and workers are sterile and may be male or female. The functional reproductives are of two types, referred to as primary and secondary, or supplementary.

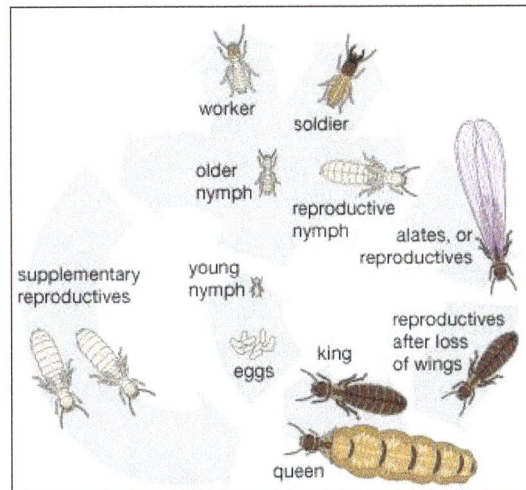

Life cycle of the termite.

Mechanisms controlling differentiation of termites into castes are not understood fully. It is known that all nymphs are genetically identical at hatching and that all could develop into any of the three major castes.

The number of individuals in each caste in a colony is closely regulated. Normally there are one pair of reproductives and a set ratio of soldiers to workers and nymphs. If members of any caste are lost, additional members of that caste develop from nymphs to restore the balance. Conversely, if overproduction of one caste occurs, selective cannibalism restores the balance.

Chemical substances such as pheromones and hormones play a role in differentiation, production, and regulation of castes. Both reproductive and soldier castes secrete a pheromone (chemical signal) that is transmitted through food sharing (trophallaxis) and grooming to other members of the colony and inhibits development of reproductives or soldiers. If the caste balance of the colony is upset, some undifferentiated nymphs do not receive the "pheromone message" and thus develop into reproductives or soldiers, thereby restoring the balance.

This inhibition theory has been confirmed by experiments with supplementary reproductive development in Kalotermes and Zootermopsis.

Pheromones may act to control caste differentiation through hormonal action, but it is not clear how this mechanism works. Activation of the corpora allata near the brain may result in release of a hormone, distinct from the juvenile hormone, that causes differentiation of a nymph into a soldier. Similarly, activity of a molt gland may be responsible for differentiation of the reproductive caste. In termites, therefore, hormones not only control molting and metamorphosis, as in other insects, but may also play a role in caste differentiation.

Colony Formation and Development

Swarming

A new termite colony normally is founded by dispersion of winged adults (alates), which usually develop in a mature colony during certain seasons of the year. After molting into winged adults, alates group themselves in special chambers near the periphery of the nest for several days or weeks. Emergence and flight of alates is usually associated with high atmospheric humidity in combination with temperature, climatic, and seasonal factors that vary with the species. In some species one emergence a year may occur; in others there may be many successive flights.

Workers prepare tunnels to the surface, open exit holes, and sometimes construct launching platforms prior to emergence of the alates. During emergence, the soldiers guard the exit holes, not only to prevent entry of enemies but also to prevent alates from re-entering the nest. At the time of emergence the alates, which normally avoid light, become attracted to it and fly out of the nest. They are weak fliers and, unless carried by the wind, descend within several hundred metres of the original colony. The flight, commonly called a nuptial or mating flight, is simply dispersal. Mating occurs after the flight. Swarming from many colonies occurs simultaneously in a given area and may be synchronized closely in areas separated by hundreds of miles. An advantage of synchronization might be intercolony mating.

Shortly after the alates land, they shed their wings, leaving only the base of the wing scale attached to the thorax. During a short courtship, in which the female raises her abdomen and emits a sex attractant, the pair moves off in tandem (pairing), with the male following closely behind the female. The couple then seek a nesting site in a crevice or dig a hole in wood or soil that has been softened by rain and seal the hole with their own fecal matter. Copulation takes place only after the establishment of this nuptial chamber. During copulation, which occurs intermittently throughout the lives of the king and queen, sperm are transferred and stored in the spermatheca of the female. Since the male has no external copulatory organ, sperm are released through a median pore on the ninth sternite, or abdominal plate.

Queen termite (genus Macrotermes) surrounded by workers.

After copulation the first batch of eggs, usually few in number, is laid. In two to five years, as the colony matures, the egg-laying capacity of the queen increases as her ovaries and fat bodies develop and her abdomen enlarges (a process called physogastry). Physogastric queens in more advanced termite families (e.g., Termitidae, especially Macrotermes and Odontotermes) may become 11 cm (4.3 inches) long. The queen becomes an "egg-laying machine" and may produce as many as 36,000 eggs a day for many years. The king is 1 to 2 cm (0.4 to 0.8 inch) long. In temperate regions egg laying stops or slows during the winter months, while in tropical species it continues throughout the year.

The first young nymphs develop into workers or pseudergates and soldiers. Only after the colonies are mature do winged adults develop. During the initial stages of colony formation, the reproductives feed the young and tend the nest. However, as the colony becomes established, the young nymphs perform these duties.

Primitive termite families have small colonies containing hundreds to thousands of individuals. More advanced families (e.g., Rhinotermitidae, Termitidae) have colonies that may number in the thousands to millions of individuals, with all members produced by the single reproductive pair. Workers and soldiers may live two to five years. The primary king and queen in higher termite families may live 60 to 70 years. The entire colony may exist for many years in species that replace the primary king and queen with secondary reproductives.

Other Colonizing Methods

Sometimes new colonies are formed by budding, the division or accidental separation of part of a colony from an original nest. When this occurs, supplementary reproductives take over as the reproductive pair. Another method of colony formation is sociotomy, or social fragmentation. In this situation, workers, soldiers, and nymphs migrate to a new nesting site, and this fragment of the original colony develops supplementary reproductives. Sometimes an original reproductive pair joins a migrating group.

Nests

Internal Features

Since termites have a soft cuticle and are easily desiccated, they live in nests that are warm, damp, dark, and sealed from the outside environment. These nests are constructed by workers or old nymphs. The high relative humidity in the interior of the nest (90 to 99 percent) probably is maintained in part by water production resulting from metabolic processes of individual termites. The temperature inside the nest generally is higher than that of the outside environment. In addition to providing an optimum microclimate, the nest provides shelter and protection against predators.

Since the anaerobic protozoans—which are necessary for cellulose digestion and live in the hindguts of primitive termites—cannot tolerate high concentrations of oxygen, these termites have developed a tolerance for high concentrations of carbon dioxide. In some species, this may be as high as 3 percent. However, ventilation must occur in the nest and is often facilitated by nest architecture. For example, the subterranean nests of Apicotermes have an elaborate system of ventilation pores. Convection currents and diffusion through the nest wall also provide ventilation in large nests.

Nest types

The family Kalotermitidae and the subfamily Termopsinae (family Hodotermitidae) make their nests in the wood on which they feed. These termites excavate irregular networks of galleries with no external openings, except the temporary ones created during swarming. The nest galleries have partitions made of fecal matter and are lined or coated with plaster made of fecal matter. The Kalotermitidae live in the sound wood of stumps and branches of trees. Examples are Neotermes tectonae, which lives in and attacks teak trees in Java, and Cryptotermes, which bores into trees and furniture in various parts of the world. The Termopsinae live in damp rotten logs. Although true wood dwellers never invade soil, and their nests have no soil connections, all other termites are basically subterranean, building their nests either in soil or with soil connections and exploiting food sources away from the nest.

Many species of Rhinotermitidae build nests in wood that is buried in damp soil and from which a diffused network of tunnels to food sources may radiate into the soil or above the ground in the form of covered runways. Other termites build a diffused subterranean nest with many chambers or pockets in soil and a network of galleries.

Many termites build discrete and concentrated nests. Some nests rise partly above the ground as mounds or hills, whereas others are totally underground or arboreal. Dirt, particles of fine clay, or chewed wood glued together with saliva or excreta are used to build nests. During nest construction a termite deposits fecal matter to cement particles in place.

Arboreal nests are ovoid structures built of "carton" (a mixture of fecal matter and wood fragments), which resembles cardboard or papier-mâché. Carton may be papery and fragile, or woody and very hard. The inside of an arboreal nest consists of horizontal layers of cells, with the queen occupying a special compartment near the centre. The nests always maintain connections with the ground through covered runways.

The large termite mounds, or hills, which are a prominent landscape feature in the tropics, may be domelike or conical. Some have chimneys and pinnacles. Longitudinal and horizontal chambers and galleries comprise the interior. Generally the outer wall is constructed of hard soil material, distinct from the internal central portion (or nursery), which is composed of softer carton material. In northern Australia Amitermes meridionalis builds wedge-shaped mounds, called compass or magnetic mounds, that are 3 to 4 metres (9.8 to 13.1 feet) high, 2.5 metres (8.1 feet) wide, and 1 metre (3.2 feet) thick at the base. The long axis is always directed north-south, and the broad side faces east-west, an orientation that probably functions to help regulate temperature. Spectacular mounds are built by fungus-growing termites in Indomalaya and Africa. Mounds of some African Macrotermes species reach a height of 8 to 9 metres (26.2 to 29.5 feet) and have pinnacles, chimneys, and ridges on their outer walls. Such mounds are built of fine particles of clay glued together by saliva to form an exceedingly hard substance. Inside the mounds are honeycomb-like structures on which the fungus grows.

Many termite nests harbour various other invertebrates as guests (e.g., beetles, flies, bugs, caterpillars, millipedes). Some, called termitophiles, are unable to survive independent of their termite hosts. True termitophiles actually have evolved with their hosts and are species specific. Some beetles and flies have developed glands that secrete substances sought and licked by the termites. The termite nest, because it provides shelter and warmth, may also be occupied by lizards, snakes, scorpions, and some birds.

A few termites, known as inquilinous species, live only in obligatory association with other termite species. Examples of such obligate relationships are Ahamitermes and Incolitermes species, which live only in the mound nests of certain Coptotermes species. In these, the galleries of guests and hosts are completely separate. Inquilinous species feed on the inner carton material of the host nests. Incolitermes, however, depend on the host species not only for food but also for exit holes from the nest during swarming when alates of the inquiline and the host emerge together. Such species' tolerance is highly unusual. Normally, different species of termites are hostile to one another, and host termites may attack inquilinous guests if partitions between galleries are broken.

Form and Function

Castes and their Roles: Reproductives

The primary reproductives in a termite colony are usually one royal pair, a king and queen. They have developed from winged forms (alates) that have flown from a parent colony and shed their wings. Because they spend time outside of the colony on the mating flight, they have hardened, pigmented bodies and large compound eyes. The primary reproductives have several important functions: reproduction, dispersal, and colony formation. In addition, during the initial stages of colony formation, the primary reproductives perform tasks later taken over by the worker caste (e.g., nest construction, housekeeping, care of young).

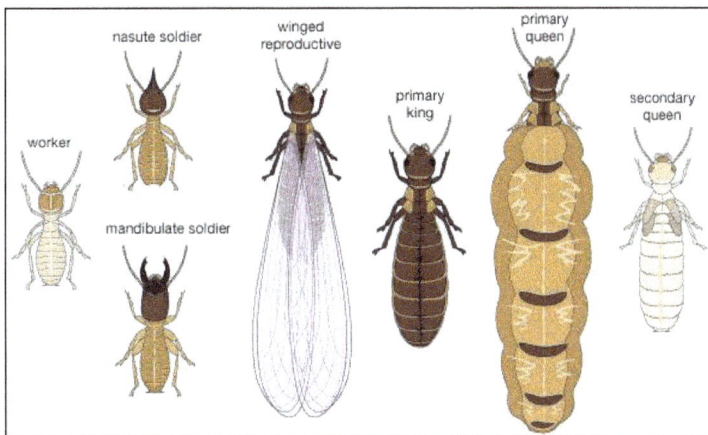

Termite castes

If the king or queen dies, it is replaced by several supplementary reproductives that are slightly pigmented and have either short wing pads (brachypterous) or none (apterous) and reduced compound eyes. These secondary reproductives, which develop from nymphs and may be called neotenics, normally are not present in a colony as long as the primary reproductives remain healthy. If a primary reproductive is lost, a neotenic achieves sexual maturity without attaining a fully winged adult stage or leaving the nest.

Workers and Soldiers

The sterile castes are the workers and soldiers. Both are wingless and usually lack eyes. Although these can be either male or female, they lack fully developed reproductive organs. In some species the workers and soldiers are dimorphic (of two sizes), with the larger individuals called major soldiers or workers and the smaller ones called minor soldiers or workers. A few species contain trimorphic soldiers. Most termite species have both soldier and worker castes.

The worker caste usually contains the greatest number of individuals in a colony. Workers are pale in colour, soft-bodied, and have hardened mandibles and mouthparts adapted for chewing. They feed all the other members of the colony (reproductives, soldiers, and young), collect food, groom other colony members, and construct and repair the nest. The worker caste is responsible for the widespread destruction termites can cause. In some primitive termite families a true worker caste is absent, and its functions are carried out by immature individuals called pseudo-workers or pseudergates, which may molt without much change in size.

The primary function of the soldier caste is defense. Since most termite soldiers are blind, they probably locate enemies through tactile and chemical means. The typical termite soldier has a large, dark, hard head. Its long powerful jaws (mandibles) may be hooked and contain teeth. The head and the mandibles are used to defend the colony against predators, usually ants. The attacking soldier makes rapid lunging movements, opening and closing its mandibles in a scissorlike action that can behead, dismember, lacerate, or grip a foe. In some soldiers (e.g., Capritermes) the mandibles are of an asymmetrical, snapping type, with the left mandible twisted and arched and the right bladelike. In defense, the mandibles lock together and release with a loud click, like the snapping of fingers. Some soldiers (e.g., Cryptotermes) use their heads, which are short and truncated in front (phragmotic), to plug the entrance holes of nests.

The higher termites (Termitidae) may supplement or replace mandibular defenses with chemical mechanisms that utilize sticky, possibly toxic, liquids secreted by either the salivary or the frontal glands. The whitish or brownish liquid becomes rubberlike after exposure to air and entangles enemies. The frontal gland of some termites (e.g., Coptotermes and Rhinotermes) occupies a large portion of the abdominal cavity and opens by means of a frontal pore (fontanelle), through which the liquid is ejected. The liquid from the frontal pore of the minor soldier of Rhinotermes flows down a groove in the elongated labrum to its hairy tip, where it volatilizes as a repellent gas.

The mandibles of soldiers with exclusively chemical defenses (Nasutitermitinae) have become reduced in size and are nonfunctional. In these, the head has become elongated into a long snout (nasus), and the frontal gland, which occupies a major portion of the head, opens at the end of the snout. These nasute soldiers can accurately fire a clear, sticky, resinous liquid for many centimetres. A few genera lack a soldier caste, and the mechanisms for defense in these groups are not known.

Nutrition

Cellulose

The food of termites is mainly cellulose, which is obtained from wood, grass, leaves, humus, manure of herbivorous animals, and materials of vegetative origin (e.g., paper, cardboard, cotton). Most lower termites and many higher ones feed on wood that is either sound or partly decayed. A few termites, known as foragers or harvesters, collect and eat grass, leaves, and straw. Many higher termites (family Termitidae) are humivores, or exclusively humus feeders.

As with other social insects, not all members of a termite colony feed directly. Because reproductives, soldiers, and young nymphs in lower families (all nymphs in Termitidae) cannot feed themselves directly, they must be fed by workers. Workers, or in families without them, the older nymphs, feed for the entire colony and transfer food to dependent castes either by mouth feeding or by anal feeding. Food transferred by mouth may consist of either pastelike regurgitated chewed wood and saliva or a clear liquid. This method is used in all termite families. During anal feeding, present only among lower termites, a pastelike liquid or droplet is discharged from the anus of the worker and licked away by the dependent castes. This liquid food, distinct from feces, consists of hindgut fluid containing protozoans, products of digestion, and wood fragments.

Cellulose digestion in lower termite families depends upon symbiotic flagellate protozoa, which live anaerobically (without oxygen) in the termite hindgut and secrete enzymes (cellulase and cellobiase) that break down cellulose into a simple sugar (glucose) and acetic acid. The termites depend entirely on protozoans for cellulose digestion and would starve without them. Newly hatched nymphs acquire protozoa from older, infected termites during anal feeding, a type of feeding necessary to lower termites that harbour protozoans.

Since the protozoans lost at the time of each molt are reacquired only through anal feeding, termites live in groups that allow contact of molting nymphs with infected, nonmolting individuals. It is possible that the necessity for transfer of protozoans was responsible for the evolution of the termite society.

Higher termites lack symbiotic protozoans, and only bacteria are present in the gut. Digestion may occur with the aid of bacterial cellulase and cellobiase enzymes, but in some species the termites themselves may secrete the enzymes.

In addition to cellulose, termites require vitamins and nitrogenous foods (e.g., proteins), which probably are supplied by fungi normally present in the decayed wood diet common to most termites. The fungi also may break down wood into components that are easily digested by termites.

Fungus Gardens

The Macrotermitinae (family Termitidae) cultivate symbiotic fungi (Termitomyces). The termites construct spongelike "fungus gardens," or combs, possibly of fecal matter rich in the carbohydrate lignin. The fungi grow on the combs, and the termites consume both fungi and combs. The fungi break down the fecal matter used to construct the combs into substances that can be reutilized by the termites. Nitrogen other than that from fungi is supplied by controlled cannibalism. The termites consume cast-off skins and dead, injured, and excess members of the colony.

Communication

Among the members of a termite colony there is continuous exchange of information, such as alarm, indication of direction and presence of a food source, and, among reproductives, calling and pairing behaviour. Information is communicated mainly by vibrations, physical contact, and chemical signals (e.g., odour). Visual cues may be used by individuals outside of the colony where light is present, but they play no role in the dark colony interior.

Many termite species leave their nests to forage for food. Workers (or older nymphs) and soldiers march in columns along the ground and carry grass, pine needles, and seeds for storage in the nest. The foraging trail between the nest and the food source may be indicated by deposits of fecal matter, covered runways over the trail, or pheromones secreted by a sternal gland as the termite drags its abdomen along the ground. The pheromone odour is detected by other termites through olfactory receptors.

Termites communicate alarm by vibrations, odour, and physical contact. Alarmed termites may tap their heads against the ground, quiver and jerk, or run in a zigzag fashion, bumping into other individuals. Although the vertical head-tapping movements produce rattling sounds audible to the human ear, termites cannot hear airborne sounds. It is the substratum vibration that they sense through the vibratory receptors located on their legs. The zigzag and horizontal jerking movements communicate alarm by contact; as an alarmed termite bumps into other termites, they, too, become alarmed. During this excitatory running, the alarmed termite leaves a scent trail, similar to the foraging trail, of pheromone that communicates direction and serves to recruit workers and soldiers to the point of disturbance.

Cockroach

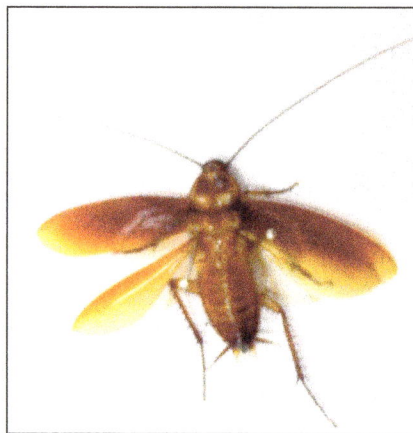

- The body of the cockroach is elongated and segmented.

- It is dark brown or reddish brown in colour.

- The exoskeleton is thick and hard made up of calcareous plates called sclerites. There are 10 segments. The segments on:

    ◦ On dorsal side (or notum) are called Tergum.

    ◦ On ventral side are called Sternum.

- The exoskeleton is coated with wax impermeable to water. It protects the body from loss of water and provides rigidity and surface for attachment of body muscles.

- The adjacent segments are joined by thin, soft and flexible arthroidal membrane.

- The body is divisible into head, thorax and abdomen.

- The cockroach has three pairs of jointed appendages and two pairs of wings.

- The fore wings are mesothoracic and are called wing covers or tegmina or elytra. They cover the hindwings and are protective in function. These are dark stiff opaque and leathery.

- The hind wings are large, thin, membranous and transparent. They are kept folded below the tegmina and are used for flying.

Mouth Parts of Cockroach

Ventrally, an opening called mouth is present on the head that remains surrounded by the mouth parts consisting of a pair of mandibles, first maxillae, labium or fused second maxillae, hypopharynx and labrum.

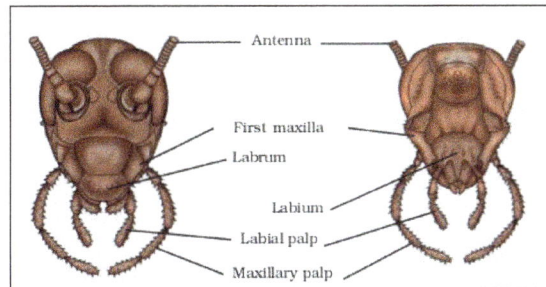

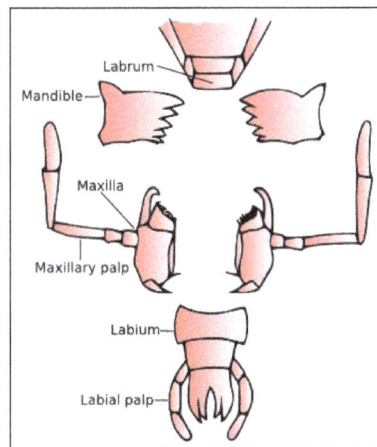

The mouth parts of the cockroach help in 'biting and chewing' its food.

Functions of the mouth parts:

- Labrum: It is the broad, flattened terminal sclerite of the dorsal side of head capsule, movably articulated to the clypeus acts as upper lip. It has epipharynx (chemoreceptors) on its inner side.

- Mandibles: Thick hard and triangular appendages beneath the labrum, on each lateral side of the mouth, which bear pointed, teeth like processes called denticles.

- First maxillae: Located on each side of the mouth next to mandibles for cutting and chewing. They also bear olfactory receptors.

- Labium: The second maxillae are fused together forming a single large structure which covers the mouth from ventral side, hence called the 'lower lip' or labium. It bears tacticle and gustatory sensory setae.

- Hypopharynx: It is a small, cylindrical mouthpart, sand witched between first maxillae and covered by labrum and labium on dorsal and ventral sides respectively. It bears several sensory setae on its free end, and the opening of common salivary duct upon its basal part.

Compound Eye

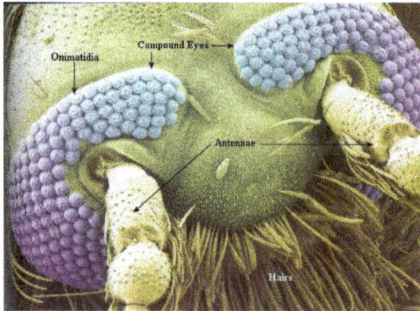

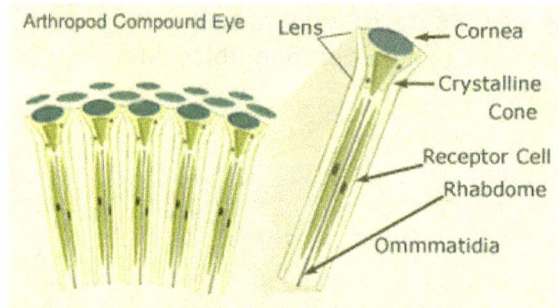

The compound eyes are situated at the dorsal surface of the head. Each eye consists of about 2000 hexagonal ommatidia (sing.: ommatidium). With the help of several ommatidia, a cockroach can receive several images of an object. This kind of vision is known as mosaic vision with more sensitivity but less resolution, being common during night (hence called nocturnal vision).

Leg of Cockroach

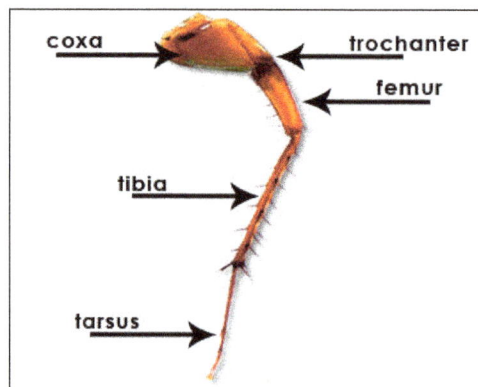

A cockroach's thorax attaches three pairs of legs. Each of the three pairs of legs is named after the region of the thorax to which it attaches:

The prothoracic legs are closest to the cockroach's head. These are the shortest legs, and they act like brakes when the roach runs. The middle legs are the mesothoracic legs. They move back and forth to either speed the roach up or slow it down.

The very long metathoracic legs are the cockroach's back legs, and they move the cockroach forward.

These three pairs of legs, are substantially different in lengths and functions, but they have the same parts and move the same way. The upper portion of the leg, called the coxa, attaches the leg to the thorax. The other parts of the leg approximate parts of a human leg:

- The trochanter acts like a knee and lets the cockroach bend its leg.

- The femur and tibia resemble thigh and shin bones.

- The segmented tarsus acts like an ankle and foot. The hook-like tarsus also helps cockroaches climb walls and walk upside down on ceilings.

Posterior Abdominal Segments

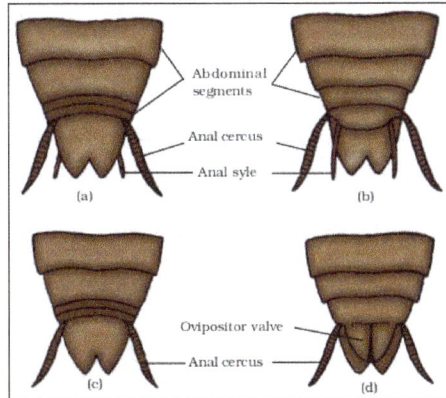

The figure shows posterior abdominal segments of cockroach
(a) Male dorsal view (b) Male ventral view
(c) Female dorsal view (d) Female ventral view.

Side View of Cockroach Showing the Location of the Systems

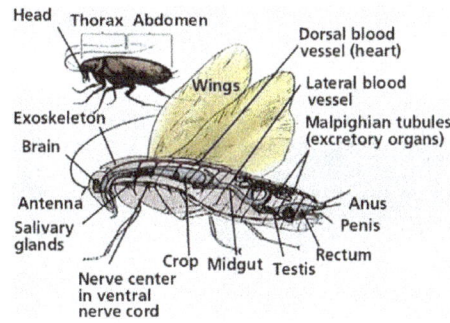

Digestive System

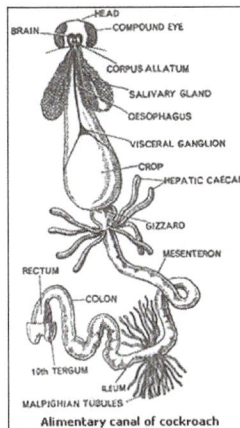

1. The alimentary canal is long and somewhat coiled divisible into three main parts namely foregut, midgut and hindgut.

2. Foregut (stomadaeum) is differentiated into five parts: Buccal chamber, pharynx, oesophagus, crop and gizzard.

3. Gizzard is muscular and internally provided with six cutical teeth which crushes the food.

4. A stomodaeal valve is present between gizzard and mesenteron.

5. Midgut (mesenteron or ventriculus) is short, tubular lined with glandular endoderm.

6. At anterior end of mesenteron there are eight blind glandular hepatic caecae which secrete digestive enzymes.

7. Hindgut (proctodaeum) comprises ileum, colon and rectum.

8. The wall of rectum is provided with six rectal papillae. They help in the absorption of water and salts.

9. Cockroach is omnivorous feeds on all sorts of organic debris.

10. The digestive enzymes of saliva are mainly zymase and amylase.

11. Most of the nutrients of food are digested in the crop.

12. Absorption of digested food takes place in mesenteron.

Nervous System

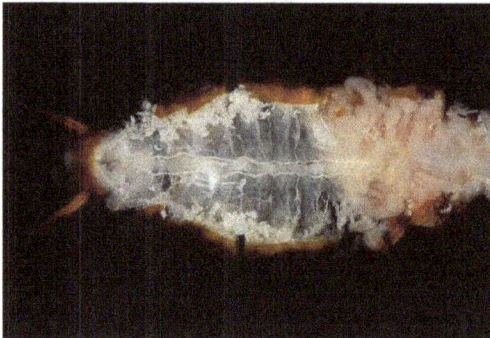

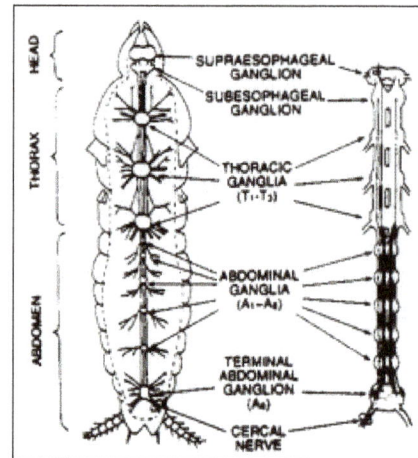

It consists of:

1. Central Nervous System.

2. Peripheral Nervous System.

3. Sympathetic or Visceral System.

   • Central Nervous System: It consists of brain or supra oesophageal ganglion. Brain gives off a pair of short, stout cords, the circumoesophageal connectives, that encircle the oesophagus and pass downwards and backwards over the suboephageal ganglion situated below oesophagus.

   From the sub esophageal ganglion passes backwards into the thorax, a double ventral nerve-cord, which bears three ganglia in the thorax and six in the abdomen.

- Peripheral Nervous System: It consists of nerves, which are given off from the ganglia so as to innervate all the parts of the body.

- Sympathetic or Stomatogastric Visceral Nervous System: It consists of a frontal ganglion, which is situated on the dorasl side of the oesophagus in the head. From this ganglion, a median unpaired recurrent nerve reaches the visceral ganglion situated on the crop. Various nerve branches are given off from the visceral ganglion.

  The frontal ganglion is jointed with the central nervous system by nerves, which connect the circumoesophageal commissures.

Sensory Structures

1. Thigmoreceptors: They are receptors of touch. Thigmoreceptors are present on body, antenna, maxillary palps and legs.

2. Olfactory receptors: They receive various smells. Olfactory receptors are present on antenna and palps.

3. Gustatory receptors: They are for sense of taste. Gustatory receptors are present on maxilla and labial palps.

4. Thermoreceptors- detect changes in temperature, present on the pads between the first four tarsals.

5. Auditory receptors- for hearing, present on the anal cerci respond to air or earth borne vibrations.

Circulatory System

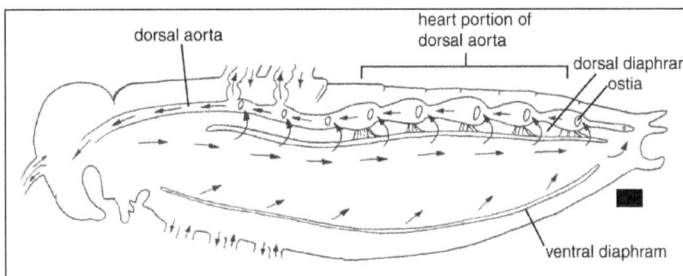

Blood vascular system is open and lacunar type. Body cavity contains blood, which bathes viscera in it therefore known as Haemocoel.

Blood vascular system consists of a tubular heart, a blood vessel called anterior aorta and a system of ill defined blood spaces or sinuses.

The Blood Sinuses

The large body cavity or haemocoel is divided by two membranous horizontal partitions, into three wide and flattened sinuses-the dorsal pericardial sinus containing the 'heart', the middle perivisceral sinus containing the gut, and the ventral perineural sinus or sternal sinus containing the nerve cord. The partition between pericardial and perivisceral sinuses is called dorsal diaphragm and between perivisceral and perineural sinuses is called ventral diaphragm. The sinuses inter-communicate by pores in the respective diaphrams. A pair of fan like, triangular alary muscles in the floor of the pericardial sinus in each segment reinforce the dorsal diaphrams by their broad bases and also connect it, by their pointed tips with the tergite of the segment.

Circulation of Haemolymph

The pumping force that propels the haemolymph is provided by the pulsations of the 'heart'. The respiratory movements of abdomen and contraction of alary muscles increase this force.

1. From the pericardial sinus, the haemolymph enters into heart through ostia.

2. The valve like ostia close, preventing back flow of haemolymph into the pericardial sinus. Therefore, some of its haemolymph is pumped into segmental vessels, while most of its poured into the head sinus through the terminally opening anterior aorta.

3. From the head sinus, the haemolymph flows backward into the thorax and abdomen. While flowing backwards from head sinus, the haemolymph remains in the ventral part due to presence of Oesophagus in dorsal part and so it fills into the perineural sinus.

4. From the perineural sinus, the haemolymph, now, flows into the perivisceral sinus through the pores of ventral diaphram in abdominal region.

5. Then from perivisceral sinus, it flows into pericardial sinus through the pores of dorsal diaphram. Then, during heart's diastole, it fills in the heart through the ostia.

Respiratory System

The respiratory system consists of a network of trachea, that open through 10 pairs of small holes called spiracles present on the lateral side of the body. Thin branching tubes (tracheal tubes sub-divided into tracheoles) carry oxygen from the air to all the parts. The opening of the spiracles is regulated by the sphincters. Exchange of gases take place at the tracheoles by diffusion.

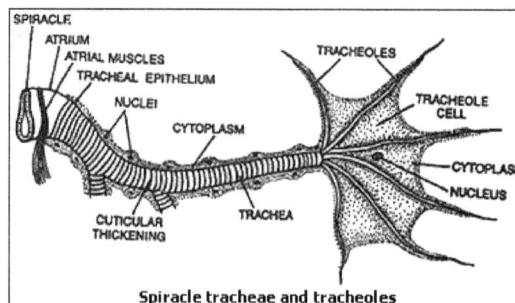

Spiracle tracheae and tracheoles

Excretory System

Excretion is performed by Malpighian tubules. Each tubule is lined by glandular and ciliated cells. They absorb nitrogenous waste products and convert them into uric acid which is excreted out through the hindgut. Therefore, this insect is called uricotelic.

In addition, the fat body, nephrocytes and urecose glands also help in excretion.

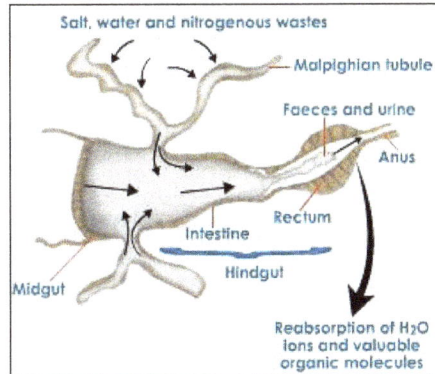

Reproductive System of Cockroach – Male

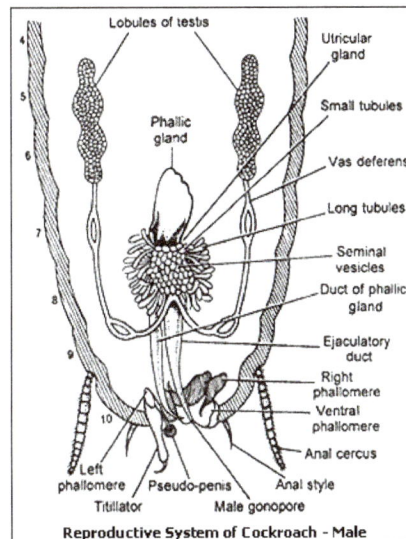

Reproductive System of Cockroach - Male

In cockroach, sexes are separate, so dioecious.

1. Testes of cockroach are located in the abdominal segments 4, 5 and 6.

2. Mushroom gland (utricular gland)consists of two types of tubules, (a) the long slender tubules the utriculi majores of peripheral tubules and (b) short tubules, the utriculi breviores, making up of the major part of the gland.

3. Small seminal vesicles are also found associated with mushroom gland.

4. All sperms of a seminal vesicle are glued together into a large bundle called spermatophore.

5. Spermatophore has three layered wall: inner layer secreted by utriculi majores; middle layer secreted by ejaculatory duct and outer layer secreted by phallic gland or conglobate gland.

6. There are three asymmetrical chitinous structures called male ganopophytes or phallomeres. Right phellomere, left phallomere (largest) and ventral phellomere (smallest).

Reproductive System of Cockroach – Female

1. Female organ consist of ovaries, oviduct, vagina, genital chamber, spermathecae, colleterial glands and female ganopophysis (ovipositor processes).

    Ovaries of cockroach are located in the abdominal segments 2 to 6. Each ovary consists of eight ovarioles.

2. Two oviducts from each side open into a common oviduct or vagina which open into genital chamber.

3. A pair of collaterial glands also open in genital chamber.

4. Genital pouch or gynatrium is divisible into a genital chamber in front and oothecal chamber behind.

5. Female genitalia consists of 3 pairs of chitinous processes hanging from the roof of oothecal chamber into its cavity.

6. Ootheca of cockroach contains sixteen fertilized eggs. Ootheca of cockroach is formed of a protein secreted by colleterial glands.

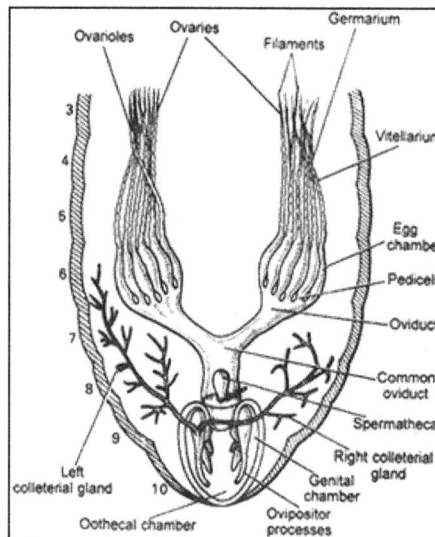

Periplaneta americana female reproductive organs in dorsal view

7. Nymph of cockroach emerge out from ootheca. A nymph resembles to adult in general structure but lacks the wings and mature reproductive organs.

8. Instar a large stage in the development of insects (larval instar, nymphal instar). Period between two moults in insects is termed stadium.

9. In periplaneta americana the nymph grows by moulting about 13 times to reach the adult from blatta orientalis moults 6 times.

10. Metamorphosis is regulated by two hormones, ecdysone secreted by prothoracic glands and juvenile hormone secreted by corpora allata.

Beetle

Beetles are the most diverse group of insects. Their order, Coleoptera (meaning "sheathed wing"), has more species in it than any other order in the entire animal kingdom. Nearly half of all described insect species are classified as beetles, and overall there are about 400,000 known species of beetles—or about one-quarter of all named species in the plant and animal kingdoms. In addition, new species are regularly discovered. Estimates put the total number of beetle species—described and undescribed—at between 5 and 8 million.

The vast numbers of beetles led to the famous quip, perhaps apocryphal, from British geneticist J. B. S. Haldane, who, when asked what one could conclude as to the nature of God from a study of his creation, replied: "An inordinate fondness for beetles". Haldane himself was a noted atheist and this quote reflects not only the vast numbers of beetles but also Haldane's skeptical perspective on natural theology.

Beetles can be found in almost all habitats, but are not known to occur in the sea or in the polar regions. They have a major impact on the ecosystem in three ways: feeding on plants and fungi, breaking down animal and plant debris, and eating other invertebrates. Certain species can be agricultural pests, for example the Colorado potato beetle (Leptinotarsa decemlineata), while other species are important controls of agricultural pests, for example the ladybirds (family Coccinellidae) consume aphids, fruit flies, thrips, and other plant-sucking insects that damage crops.

The study of beetles is called coleopterology; its practitioners are coleopterists. There is a thriving industry in the collection of wild caught species by amateur and professional collectors.

Anatomy

The anatomy of beetles is quite uniform. Beetles are generally characterized by a particularly hard exoskeleton, and the hard wing-cases (elytra) that tend to cover the hind part of the body and protect the second wings, the alae. The elytra are not used in flying, but generally must be raised in order to move the hind wings. In some cases, the ability to fly has been lost, characteristically in families such as Carabidae (ground beetles) and Curculionidae (snout beetles and true weevils). After landing, the hind wings are folded below the elytra.

In a few families, both the ability to fly and the wing-cases are absent, with the best known example being the "glowworms" of the family Phengodidae, in which the females are larviform throughout their lives.

The bodies of beetles are divided into three sections, the head, the thorax, and the abdomen, and these in themselves may be composed of several further segments.

The eyes are compound, and may display some remarkable adaptability, as in the case of the Whirligig beetles (family Gyrinidae), in which the eyes are split to allow a view both above and below the waterline. The dorsal appendage aids the beetle in stalking prey.

Like all insects, antennae and legs are both jointed.

Oxygen is taken in via a tracheal system: this takes air in through a series of tubes along the body, which is then taken into increasingly finer fibers. Pumping movements of the body force the air through the system. Beetles have hemolymph instead of blood, and the open circulatory system of the beetle is powered by a tube-like heart attached to the top inside of the thorax.

Development

Beetles are endopterygotes—a superorder of insects of the subclass Pterygota that go through distinctive larval, pupal, and adult stages, or complete metamorphosis. The larva of a beetle is often called a grub and represents the principal feeding stage of the life cycle.

The eggs of beetles are minute, but may be brightly colored. They are laid in clumps and there may be from several dozen to several thousand eggs laid by a single female.

Once the egg hatches, the larvae tend to feed voraciously, whether out in the open such as with ladybird larvae, or within plants such as with leaf beetle larvae.

As with lepidoptera, beetle larvae pupate for a period, and from the pupa emerges a fully formed beetle or imago.

In some cases, there are several transitory larvae stages; this is known as hypermetamorphosis. Examples include the blister beetles (family Meloidae).

The larval period of beetles varies between species, but can be as long as several years. Adults have an extremely variable lifespan of weeks to years.

Larva of the cockchafer

Reproduction and Parental Care

Beetles may display some intricate behavior when mating. Smell is thought to be important in the location of a mate.

Conflict can play a part in the mating rituals, for example, in species such as burying beetles (genus Nicrophorus) where localized conflicts between males and females rage until only one of each is left, thus ensuring reproduction by the strongest and fittest. Many beetles are territorial and will fiercely defend their small patch of territory from intruding males.

Pairing is generally short, but in some cases will last for several hours. During pairing, sperm cells are transferred to the female to fertilize the egg.

Parental care between species varies widely, ranging from the simple laying of eggs under a leaf to scarab beetles, which construct impressive underground structures complete with a supply of dung to house and feed their young.

There are other notable ways of caring for the eggs and young, such as those employed by leaf rollers, who bite sections of leaf causing it to curl inwards and then lay the eggs, thus protected, inside.

Striped love beetle Eudicella gralli from the forests of Central Africa.
The iridescent wing cases are used in marriage ceremonies.

Diet and Behavior

There are few things that a beetle somewhere will not eat. Even inorganic matter may be consumed.

Some beetles are highly specialized in their diet; for example, the Colorado potato beetle (Leptinotarsa decemlineata) almost entirely colonizes plants of the potato family (Solanaceae). Others are generalists, eating both plants and animals. Ground beetles (family Carabidae) and rove beetles (family Staphylinidae) are entirely carnivorous and will catch and consume small prey such as earthworms and snails.

Decaying organic matter is a primary diet for many species. This can range from dung, which is consumed by coprophagous species, such as the scarab beetles (family Scarabaeidae), to dead animals, which are eaten by necrophagous species, such as the carrion beetles (family Silphidae).

Various techniques are employed by many species for retaining both air and water supplies. For example, predaceous diving beetles (family Dytiscidae) employ a technique of retaining air, when diving, between the abdomen and the elytra.

Beetles and larvae have a variety of strategies for avoiding being eaten. Many employ simple camouflage to avoid being spotted by predators. These include the leaf beetles (family Chysomelidae) that have a green coloring very similar to their habitat on tree leaves. A number of longhorn beetles (family Cerambycidae) bear a striking resemblance to wasps, thus benefitting from a measure of protection. Large ground beetles by contrast will tend to go on the attack, using their strong mandibles to forcibly persuade a predator to seek out easier prey. Many species, including lady beetles and blister beetles, can secrete poisonous substances to make them unpalatable.

Evolutionary History and Classification

Beetles entered the fossil record during the Lower Permian, about 265 million years ago.

The four extant (living) suborders of beetle are:

- Polyphaga, the largest suborder, contains more than 300,000 described species in more than 170 families, including rove beetles (Staphylinidae), scarab beetles (Scarabaeidae), blister beetles (Meloidae), stag beetles (Lucanidae), and true weevils (Curculionidae). These beetles can be identified by the cervical sclerites (hardened parts of the head used as points of attachment for muscles) absent in the other suborders.

- Adephaga contains about 10 families of predatory beetles, includes ground beetles (Carabidae), predacious diving beetles (Dytiscidae), and whirligig beetles (Gyrinidae). In these beetles, the testes are tubular and the first abdominal sternum (a plate of the exoskeleton) is divided by the hind coxae (the basal joints of the beetle's legs).

- Archostemata contains four families of mainly wood-eating beetles, including reticulated beetles (Cupedidae) and telephone-pole beetles (Micromalthidae).

- Myxophaga contains about 100 described species in four families, mostly very small, including skiff beetles (Hydroscaphidae) and minute bog beetles (Sphaeriusidae).

These suborders diverged in the Permian and Triassic. Their phylogenetic relationship is uncertain, with the most popular hypothesis being that Polyphaga and Myxophaga are most closely related, with Adephaga an outgroup to those two, and Archostemata an outgroup to the other three.

The extraordinary number of beetle species poses special problems for classification, with some families consisting of thousands of species and needing further division into subfamilies and tribes.

Impact on Humans

Pests

There are several agricultural and household pests represented by the order. These include:

- The Colorado potato beetle (Leptinotarsa decemlineata) is a notorious pest of potato plants. Adults mate before over-wintering deep in the soil, so that when they emerge the following spring females can lay eggs immediately once a suitable host plant has been found. As well as potatoes, hosts can be a number of plants from the potato family (Solanaceae), such as nightshade, tomato, aubergine, and capsicum. Crops are destroyed and the beetle can only be treated by employing expensive pesticides, many of which it has begun to develop immunity to.

- The elm bark beetles, Hylurgopinus rufipes, elm leaf beetle Pyrrhalta luteola. and Scolytus multistriatus (in the family Scolytidae) attack elm trees. They are important elm pests because they carry Dutch elm disease (the fungus Ophiostoma ulmi) as they move from infected breeding sites to feed on healthy elm trees. The spread of the fungus by the beetle has led to the devastastation of elm trees in many parts of the Northern Hemisphere, notably North America and Europe.

- The death watch beetle (Xestobium rufovillosum) is of some considerable importance as a pest of wooden structures in older buildings in Great Britain. It attacks hardwoods, such as oak and chestnut, and always where some fungal decay has taken or is taking place. It is thought that the actual introduction of the pest into buildings takes place at the time of construction.

- Asian long-horned beetle.

- Citrus long-horned beetle.

Damage to beans by larvae of the common
bean weevil, Acanthoscelides obtectus

Beneficial Beetles

- The larvae of lady beetles (family Coccinellidae) are often found in aphid colonies, consuming these agricultural pests. While both adult and larval lady beetles found on crops prefer aphids, they will, if aphids are scarce, use food from other sources, such as small caterpillars, young plant bugs, aphid honeydew, and plant nectar.

- Large ground beetles (family Carabidae) are predators of caterpillars and, on occasion, adult weevils, which are also significant agricultural pests. Smaller species of ground beetles attack eggs, small caterpillars, and other pest insects.

To foster and provide cover for beneficial beetles, some farmers introduce beetle banks (a strip of grass or perennials that provide habitat for insects hostile to pests).

Scarab Beetles in Egyptian Culture

Several species of the dung beetles, most notably the Scarabaeus sacer (often referred to as "scarab"), enjoyed a sacred status among the Egyptians, as the creature was likened to the god Khepri. Some scholars suggested that the people's practice of making mummies was inspired by the brooding process of the beetle.

Many thousands of amulets and stamp seals have been excavated that depict the scarab. In many artifacts, the scarab is depicted pushing the sun along its course in the sky. Scarab amulets were often placed over the heart of the mummified deceased. The amulets were often inscribed with a spell from the Book of the Dead which entreated the heart to, "do not stand as a witness against me."

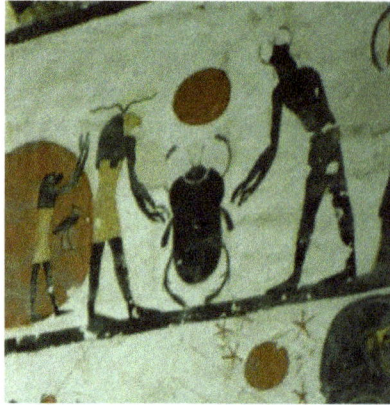

Ancient Egyptian scene depicting a scarab beetle

Aquatic Insects

Insects are invertebrates (animals without a backbone) that are part of the larger group of animals called arthropods. Arthropod means "joint footed." That name was given to these animals because all of the arthropods have legs with joints that are something like our elbows and knees. Some other arthropod relatives of insects are crayfish, crabs, lobsters, millipedes, centipedes, scorpions, spiders, and ticks. Most insects are terrestrial (live on land), and are found in places such as trees, shrubs, flowers, rocks, logs, soil, buildings, and especially our gardens. Everyone is familiar with common terrestrial insects such as butterflies, moths, beetles, ants, bees, wasps, grasshoppers, crickets, cockroaches, and flies.

There are also many kinds of insects that live in the water. These are called aquatic insects, and they are often not seen unless you explore places such as puddles, ponds, lakes, ditches, streams, and lakes. There are many different kinds of aquatic insects and almost every type of freshwater environment will have some kind of aquatic insect living in it.

Structure and Appearance

Like their other arthropod relatives, insects have their skeleton on the outside of their body (exoskeleton). This outside skeleton is thick, and often hard, so that it can protect the insect, much like our skin protects us. However, an insect's exoskeleton does not grow along with the insect. As an insect grows it must shed its skin and grow a new one in order to get larger.

There are several features of insects' bodies that make them different from the other arthropods. The body of an insect is made up of three sections. The head is at the front end of the body. The thorax is in the middle of the body and is usually larger than the head. The abdomen is the rear section of the body and is usually as long, or longer, than the head and thorax together. You can usually see individual segments (up to 10) in the abdomen, but the individual segments in the head and thorax are usually fused together and cannot be easily distinguished.

The head contains structures for eating and sensing the world that an insect lives in. Insects have several different mouthparts that are specialized for tasting, obtaining, and breaking up food.

There are two antennae (feelers) on the head, one on each side. There are usually two large compound eyes, which contain thousands of small individual eye cells.

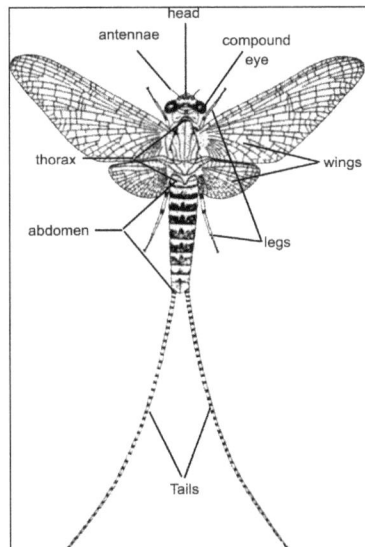

Major morphological features

The thorax contains the structures that insects use to move around. There are six legs, arranged with three on each side of the body. Some of the legs may be constructed for special movements or purposes, such as running, jumping, digging, or catching food. Most insects have four wings on the thorax. Insects are the only arthropods that can fly, which is the main reason they are more widespread and more diverse than any of the other arthropods.

The abdomen does not have many structures on the outside, except for some structures at the end that are specialized for mating and laying eggs. Many insects have two tails on the rear of the abdomen, which they often use to sense vibrations.

The distinguishing features described above apply to adult insects. For many aquatic insects, only the immature stages (babies) live in the water. Sometimes the immature stages do not have the same features as adult insects. Immature insects never have working wings, and some may not have compound eyes, jointed legs, or a distinguishable thorax section. Some immature insects look more like worms than insects. However, wormlike immature insects will always have an obvious head, or at least some noticeable structures sticking out from the head, such as mouthparts or antennae, while worms never have a head or head structures.

How do they Breathe?

Insects that live on land breathe air through holes in their bodies called spiracles. This would not work very well for insects that live in the water, so aquatic insects have special adaptations for breathing in the water without drowning. The most common way for aquatic insects to breathe effectively underwater is to use oxygen that is dissolved in the water rather than oxygen that is a gas in the air. Many aquatic insects, especially during their immature stages, have gills similar to fish for obtaining dissolved oxygen. The gills of aquatic insects are located on the outside of their body in various locations. Gills come in various shapes, but many are flat oval plates or tufts of small filaments. Other aquatic insects have a soft flexible external skeleton that simply allows dissolved oxygen to pass from the water into their body all over their body surface.

Some kinds still use the holes in their bodies to get oxygen from the air. They just keep the holes shut while they are underwater, and only open them when they come to the surface to take in a breath of air. Some kinds take a bubble of air underwater and breathe out of the bubble, which allows them to stay underwater longer. This is comparable to SCUBA diving. A few kinds of aquatic insects have their spiracles on the end of a long tube at the end of their abdomen. They keep their body underwater and just stick their breathing tube up to the surface to get air, much like snorkeling.

What do they Eat?

The foods of aquatic insects are just as diverse as the habitats they live in. Although individual kinds of aquatic insects may only eat one type of food, all organic material in the water, living and dead, is eaten by some kind of aquatic insect. Scientists have found it informative to categorize aquatic insects according to how they obtain their food for studying the ecology of freshwater ecosystems. These categories are called functional feeding groups.

Scrapers have special mouthparts that remove algae growing on the surface of rocks or other solid objects. These mouthparts work like a sharp blade to remove the outermost layer of algae, which is attached very tightly but is very nutritious for those insects equipped to remove it.

Collectors acquire small pieces of decaying plant material (detritus). Some kinds use long hairs on their head or legs or silk nets to filter these small particles out of the water. Other kinds of collectors use their mouthparts to gather fine particles lying on the bottom and shove this material into their mouths.

Shredders have mouthparts that are designed to nibble off pieces of soft vegetation, such as leaves, flowers, or twigs, and grind up this material. Most aquatic insects shred pieces of vegetation that have dropped off of plants and are decaying. Most of this material comes from trees and shrubs that grow on land at the edge of the water. Only a few kinds of aquatic insects feed on parts of live plants that grow under the water.

Predators feed on other animals that are alive. Predators often have special structures for catching and subduing their prey, such as strong jaws with teeth, a sharp beak, or spiny legs. Predators eat other invertebrates most of the time, but some are large and strong enough to catch small vertebrates, such as fish and tadpoles.

How do they Grow?

Insects, like all arthropods, must shed their protective external skeleton periodically in order to get larger. Before they shed their old skin, they grow a new one underneath. The new larger skin is soft and folded, so that it can fit inside the old smaller skin. They also absorb much of the old skin and recycle the materials it contained. To shed the old skin, aquatic insects gulp water to make themselves larger, and they push from the inside on the top of the thorax. The old exoskeleton is thinner there, so it splits and allows the insect to climb out. After getting out of the old skin, the insect continues to gulp water to make itself larger by stretching out the wrinkles in the new soft skin. After a few hours, or in some cases a few days, the new skin hardens into another protective exoskeleton. Different kinds of aquatic insects shed their skin anywhere from three to 45 times. This is a dangerous time in the life of an aquatic insect, and many of them die while they are waiting for their new skin to harden. The new soft exoskeleton is subject to damage, and it does not allow the insect to move and hide very well.

As insects grow from eggs into adults, they go through a series of developmental changes that are called metamorphoses. There are two basic types of metamorphosis in aquatic insects. It is important to know about these two types because metamorphosis is used to help identify immature insects and it explains a lot of the biological activities they engage in during their lives.

One type of metamorphosis is incomplete (also called hemimetabolous). Aquatic insects with incomplete metamorphosis emerge from the egg looking a lot like miniature versions of the adults, minus wings. They have compound eyes on the head and jointed legs on the thorax. The wings develop in projections on the thorax (wing pads) and get a little larger each time the insect sheds its skin. Scientists have begun to use the term larvae for the immature stages of aquatic insects with incomplete metamorphosis, but for many years they were called nymphs or naiads.

The other type of metamorphosis is complete (also called holometabolous). Immature aquatic insects with complete metamorphosis do not look anything like the adults they turn into. At most they have only a simple eyespot, or a small cluster of eyespots, on the head, and some have no eyes at all. They may or may not have jointed legs on the thorax. The wings develop inside the body, so no wing pads are visible until just before the insect becomes an adult. Insects with complete metamorphosis go into an inactive stage before they become adults. During the inactive stage is when compound eyes, jointed legs, and wing pads first show up. The active immature stages are called larvae, and the inactive stage is called a pupa. Some aquatic insects crawl out of the water for the pupa stage, while others spend this stage in the water.

How do they Reproduce?

Only adult insects are capable of reproducing, and most aquatic insects spend their adult stage out of the water. After mating on land, females return to the water to deposity her eggs. Eggs are usually stuck on solid objects under water, but a few kinds deposit the eggs on trees or rocks above the water.

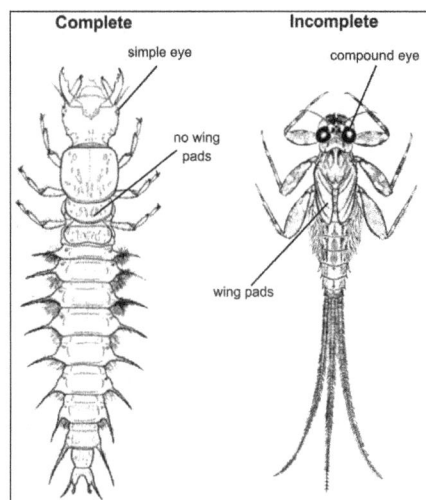

Two types of metamorphosis

Eggs usually hatch within a few days or weeks, but some may be programmed to not hatch for many months. A delay in hatching allows an aquatic insect to live in habitats that are too hot, too cold, or dry up during part of the year.

Different kinds of aquatic insects require anywhere from a few weeks to several years to develop into adults. Most grow and develop steadily, but some may go into an inactive state to endure harsh environmental conditions. It is most common for aquatic insects to produce one generation per year, with the adults emerging during the warm months. Some of the larger kinds, or those living in cold environments, require two to three years to develop from egg to adult. Some kinds with short developmental times may produce more than one generation per year. In southern latitudes of the United States, some kinds reproduce continuously throughout the year.

Adult aquatic insects usually live a few days to a few weeks. The extremes of adult life span range from a few minutes (some mayflies) to two or three years (some water beetles).

Where do they Live?

One of the most amazing things about aquatic insects is the diversity of habitats where they live. There is no body of water that is too small, too large, too cold, too hot, too muddy, with oxygen too low, with currents too fast, or even with too much pollution for some kind of aquatic insect to live there. About the only restriction to where they live is that they do not usually inhabit the salty water of marine environments, such as oceans and bays. However, there are even a couple of unusual aquatic insects that live on coral reefs and in tide pools of marine environments. Estuaries, where the fresh water of rivers mixes with the salt water of oceans, are home to quite a few kinds of aquatic insects. Anyone who has been to the beach knows about the kinds of mosquitoes that breed in the salt marshes near the beach.

Not all kinds of aquatic insects live in all types of freshwater habitats. The most favorable habitats, and the ones where you can collect the most kinds of aquatic insects, are the edges of ponds and lakes and the sections of streams and rivers where the water is flowing fast enough to splash (riffles). In both standing and flowing freshwater habitats, the most different kinds of aquatic insects will be found in water that is less than three feet deep and can be easily waded.

Aquatic insects have a variety of special adaptations for moving around or staying in one place within their habitat. Some are agile swimmers by means of streamlined bodies with long legs or tails, while others climb around on aquatic plants by means of long thin bodies. Some sprawl on top of soft mucky bottom without sinking in because their bodies are flat and their legs extend out from the sides. Others are able to burrow down into soft mucky bottom because they have special structures on their bodies, such as legs that look like shovels or points projecting in front of their heads. Still others can cling to rocks and logs in very swift current because their bodies are very flat and the current just passes over them without knocking them off. Other clingers stay put by using special suckers or by gluing themselves down with sticky silk that they produce. Lastly, many aquatic insects like to crawl around in the tiny spaces among rocks, sticks, and dead leaves.

Because aquatic insects are small and highly specialized, different kinds are often found in small areas with similar features, which are called microhabitats. Examples of microhabitats where you will probably find different aquatic insects are: cobble rocks (about the size of your fist or head), gravel, sand, muck, accumulations of dead leaves and twigs, live plants, and grasses and tree roots that extend into the water from land. Different microhabitats, with different aquatic insects living in them, occur very close together, perhaps within one step of each other.

Aquatic insects even live in temporary habitats, such as small streams or ponds that dry up in the summer. If they are adults, they can simply fly to another place with water. Some immature aquatic insects that cannot yet fly will burrow down into the bottom where it is damp and go into an inactive state, something like animals hibernating over winter. However, most aquatic insects that live in temporary habitats are "programmed" to stay in their eggs, where they are protected, until the time of year when water is present.

Major Groups of Aquatic Insects

There are so many different kinds of aquatic insects, it is difficult to appreciate their biological diversity without considering some of the individual kinds. The following section provides a brief summary of the eight major groups

Mayflies (Ephemeroptera)

Larvae of mayflies live in a wide variety of flowing and standing waters. Most of them eat plant material, either by scraping algae or collecting small pieces of detritus from the bottom. Larvae breathe dissolved oxygen by means of gills on the abdomen. They have incomplete metamorphosis. Most mayflies are sensitive to pollution, although there are a couple of exceptions. The most unusual feature of mayflies is that the adults only live a few hours and never eat.

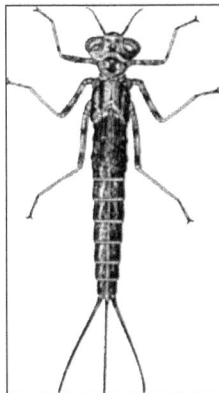

Dragonflies and Damselflies (Odonata)

Larvae of dragonflies and damselflies are most common in standing or slow-moving waters. All of them are predators. Larvae breathe dissolved oxygen with gills, which are located either inside the rear portion of the abdomen (dragonflies) or on the end of the abdomen (damselflies). They have incomplete metamorphosis. Many kinds are fairly tolerant of pollution, but some kinds only live in unique habitats, such as bogs high in the mountains. The most unusual feature of this group is the way the larvae catch their food with an elbowed lower lip, which they can shoot out in front of the head.

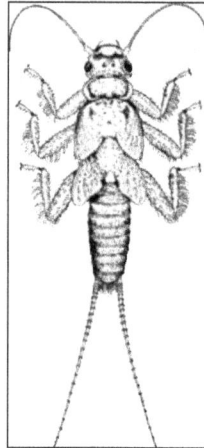

Stoneflies (Plecoptera)

Larvae of stoneflies live only in flowing waters, often cool, swift streams with high dissolved oxygen. Some feed on plant material, either by shredding dead leaves and other large pieces of detritus, while others are predators. Larvae breathe dissolved oxygen. Some have gills on their thorax, but others just obtain dissolved oxygen all over their body. They have incomplete metamorphosis. Almost all of the stoneflies are sensitive to pollution. The most unusual feature of this group is that some kinds are programmed to emerge only during the coldest months; hence, they are called the winter stoneflies.

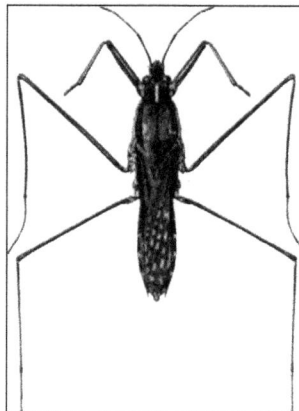

True Bugs (Hemiptera)

Most of the true bugs live on land, but the aquatic kinds are most common in the shallow areas around the edge of standing waters. Both the adults and the larvae of the aquatic kinds live in the water. Both stages are usually found on submerged aquatic plants. Almost all of them are predators. They breathe oxygen from the air, either by taking a bubble underwater or by sticking a breathing tube up into the air. They have incomplete metamorphosis. Most kinds are tolerant of pollution. The most unusual feature of this group is the way they kill and eat their prey. True bugs have a sharp beak that they stick into the body of their prey, and then they pump in poison to kill their prey, after which they suck out the body fluids. Some of the larger kinds feed on small fish and tadpoles.

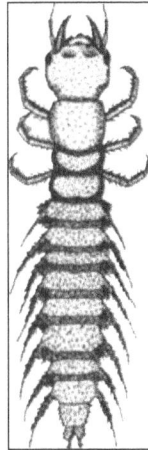

Dobsonflies and Alderflies (Megaloptera)

Larvae of different kinds live in flowing or standing waters. They are all predators. They breathe dissolved oxygen by means of gills and their overall body surface. They have complete metamorphosis. Mature larvae leave the water and dig out a protected space under a rock or log for the pupa stage. Different kinds are either sensitive or tolerant to pollution. Larvae of some of the larger kinds are called hellgrammites, which are popular as live bait for smallmouth bass and other warm-water fish species.

Water Beetles (Coleoptera)

There are more species of beetles than any other insects, but most of them live on land. Most of the water beetles are more common in standing or slowmoving waters, but a few kinds are only found in swiftly flowing waters. Both the adults and the larvae of the aquatic kinds live in the water. Water beetles feed in different ways, primarily by preying on other animals, scraping algae, or collecting small particles of detritus from the bottom. All of the adults breathe air by taking a bubble underwater, while most of the larvae breathe dissolved oxygen by a combination of gills and their overall body surface. They have complete metamorphosis and leave the water for the pupa stage. Water beetles range from sensitive to somewhat tolerant of pollution. The most unusual feature of water beetles is that some of the adults live for several years.

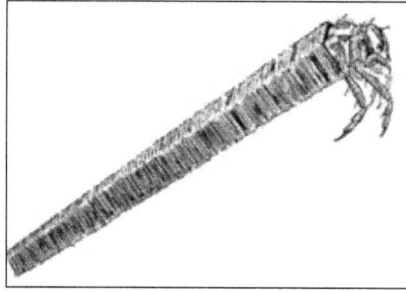

Caddisflies (Trichoptera)

Larvae of different caddisflies live in a wide variety of flowing and standing waters. They also have a wide range of feeding habits, including scraping algae, collecting fine particles of detritus from the bottom or from the water, shredding dead leaves, and preying on other invertebrates. They breathe dissolved oxygen by means of gills and their overall body surface. Caddisflies have complete metamorphosis and remain in the water for the pupa stage. Most kinds are sensitive to pollution, but a few kinds are somewhat tolerant of moderate levels of pollution. The most distinctive feature of caddisflies is their ability to spin silk out of their lower lip. They use this material to glue together stones or pieces of vegetation into a small house for their protection during the larva and pupa stages. Some also use strands of silk to make a net for filtering particles of food from the water.

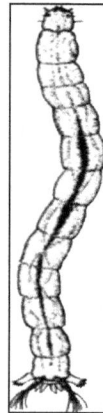

True Flies (Diptera)

This group has more kinds on land, but there are also many aquatic kinds. They have a wide range of feeding habits, including scraping algae, collecting fine particles of detritus from the bottom or from the water, shredding dead leaves, and preying on other invertebrates. They breathe dissolved oxygen by means of gills and their overall body surface. True flies have complete metamorphosis and remain in the water for the pupa stage. The most distinctive feature of this group is their ecological diversity. Some kinds live in the cleanest habitats (e.g., swift, cool, mountain streams), while others live in some of the harshest natural habitats on the earth (e.g., arctic tundra ponds, geothermal springs, alkaline lakes, mucky swamps). They have equally diverse responses to pollution, with some kinds being exceptionally sensitive, while other kinds endure the worst imaginable water quality (e.g., raw sewage or acid mine drainage).

Entomology

Entomology is the branch of zoology dealing with the scientific study of insects. The zoological categories of genetics, taxonomy, morphology, physiology, behaviour, and ecology are included in this field of study. Also included are the applied aspects of economic entomology, which encompasses the harmful and beneficial impact of insects on humans and their activities. Entomology also plays an important role in studies of biodiversity and assessment of environmental quality.

Applied Entomology

Many entomologists are employed in the study of insects that are directly beneficial or harmful to humans. Beneficial insects include those that are pollinators of agricultural crops and ornamental flowers and those imported or otherwise vital as biological control agents. The study of beneficial insects is primarily focused on their ecology and life habits, the primary concern being the understanding of how to raise them and make them more productive, or protect them from human disturbance if they are native species such as wild bees.

Conversely, much of the study of insects (and related arthropods) that directly harm human beings or their domestic animals, called medical entomology and veterinary entomology, is focused on their physiology, with the goal of developing insect controls that are effective, while minimizing undesirable side effects. For instance, many types of insecticides have been developed that target unique aspects of insect physiology and are thus considered harmless to other kinds of animals. A risk to this approach is that insecticides can also kill beneficial insects. Considerable recent effort has gone into finding biological controls that are species-specific, such as species-specific parasites and diseases, as well as genetic controls, such as the introduction of sterile insects into a population. The combination of taking into account all aspects of insect biology, available control measures, economics, and environmental considerations is known as integrated pest management.

A few insects, chiefly blood-sucking Diptera, are vectors for a wide range of deadly diseases. Mosquitoes are especially important disease vectors, with the genus Anopheles the principle vector of malaria, Aedes aegypti the main vector of yellow fever and dengue, and other Aedes spp. carrying the causal agents of various types of encephalitis. Other well-known vectors include the tsetse fly (genus Glossina transmits protozoan agents of African sleeping sickness), black flies (spread the parasitic roundworm Onchocerca volvulus, cause of onchoceriasis), and sand flies (genus Phlebotomus are vectors of bacteria that causes Carrion's disease, and sand flies are also agent of protozoans Leishmania spp. that cause Leishmaniasis).

Forensic entomology specializes in the study of insect ecology for use in the legal system, as knowledge of insect behavior can yield useful information about crimes. For example, the approximate time of death or whether or not a victim was alive during a fire may be determined by using facts such as at what stage of life is an insect found at the scene.

Taxonomic Specialization

Given the vast number and diversity of insects, many entomologists specialize in a single order or even a family of insects. A number of these subspecialties are given their own informal names, typically (but not always) derived from the scientific name of the group:

- Apiology (or melittology) - (study of) bees.

- Coleopterology - beetles.

- Dipterology - flies.

- Heteropterology - true bugs.

- Lepidopterology - moths and butterflies.

- Myrmecology - ants.

- Orthopterology - grasshoppers, crickets, etc.

- Trichopterology - caddis flies.

Identification of Insects

Insects other than Lepidoptera are typically identifiable only through the use of identification keys and monographs. Because the class Insecta contains a very large number of species, and the characters separating them are unfamiliar and often subtle (or invisible without a microscope), this is often very difficult even for a specialist.

Insect identification is an increasingly common hobby, with butterflies and dragonflies being the most popular.

Insect Life Stages and Development

Insect growth and development varies with three main types of development: ametabolous, hemimetabolous and holometabolous. Insect growth is discontinuous as it is limited by the relatively rigid skin, so growth progresses by moulting the skin. Each immature period between moults is referred to as an instar or stadium.

Ametabolous Development

In the basal (= 'primitive') orders Archaeognatha (bristletails) and Zygentoma (silverfish) growth is indeterminate (individuals continue to moult until they die), although they do increase in size when they reach adulthood. In ametabolous insects, the larvae resemble adults but lack genitalia.

Hemimetabolous Development

Hemimetabolous or incomplete development involves repeated stages of moulting and growth, where the immatures (= nymphs) and adults are similar in morphology, but the nymphs lack wings and genitalia. The wings are developed externally, with nymphs having wing buds, which explains why such insects are called exopterygotes. There are five nymphal stages.

Hemimetabolous orders include the Ephemeroptera (mayflies), Odonata (dragonflies), and the orders of the Polyneoptera (= Orthopteroid orders) – Plecoptera (stoneflies), Dermaptera (earwigs), Embioptera (webspinners), Blattodea (cockroaches & termites), Mantodea (praying mantis), Phasmatodea (stick insects) and Orthoptera (grasshoppers & crickets); and Paraneoptera (= Hemipteroid orders) – Psocodea (bark lice, true lice), Thysanoptera (thrips) and Hemiptera (bugs, cicadas, scales & others).

Holometabolous Development

The majority of insects undergo holometabolous or complete metamorphosis, where there is a major reorganisation of the body. This results in very different bodies of the immatures (= larvae) and adults of the same species (e.g. caterpillars and butterflies). This dramatic metamorphosis occurs during the inbetween pupal stage. The wings are developed inside the body hence the name endopterygotes. Holometabolous development occurs in the 'Big 4' insect orders – Coleoptera (beetles), Diptera (flies), Hymenoptera (bees & wasps) and Lepidoptera (butterflies & moths), as well as Neuroptera (lacewings), Siphonaptera (fleas) Trichoptera (caddisflies), Raphidioptera (snakeflies), Megaloptera (dobsonflies), Mecoptera (scorpionflies) and Strepsiptera (strepsipterans).

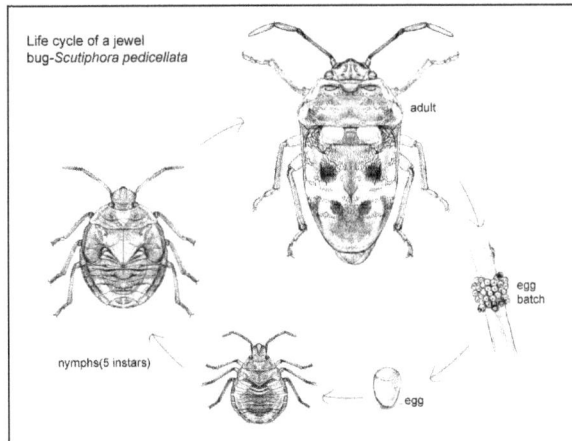

Life cycle of a jewel bug-*Scutiphora pedicellata*

adult

egg batch

nymphs(5 instars)

egg

Hemimetabolous

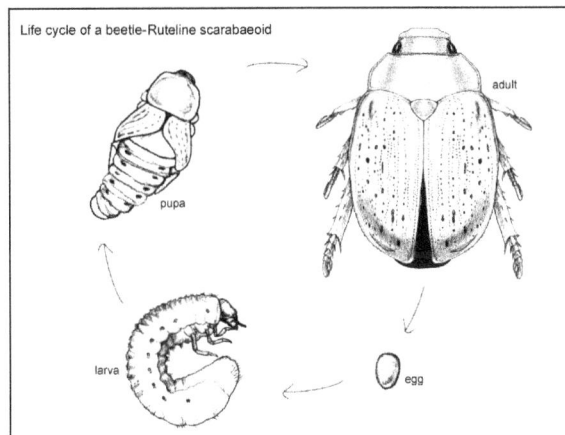

Life cycle of a beetle-Ruteline scarabaeoid

adult

pupa

larva

egg

Holometabolous

References

- Insect, animal: britannica.com, Retrieved 4 April, 2019

- Taxonomic-principles: discoverlife.org, Retrieved 12 July, 2019

- Apterygote, animal: britannica.com, Retrieved 15 February, 2019

- Bristletails, almanacs-transcripts-and-maps, science: encyclopedia.com, Retrieved January 18, 2019

- All-about-jumping-bristletails: welcomewildlife.com, Retrieved 28 May, 2019

- Silverfish: almanac.com, Retrieved 8 February, 2019

- Mayfly, animal: britannica.com, Retrieved 13 June, 2019

- Damselfly: newworldencyclopedia.org, Retrieved 3 March, 2019

- Neoptera: newworldencyclopedia.org, Retrieved 9 June, 2019

- Termite, animal: britannica.com, Retrieved 19 March, 2019

- Dunkle, Sidney W. (2000). Dragonflies Through Binoculars: a field guide to the dragonflies of North America. Oxford University Press. ISBN 978-0-19-511268-9

- Morphology-and-anatomy-of-cockroach: biology4isc.weebly.com, Retrieved 2 July, 2019

- Beetle: newworldencyclopedia.org, Retrieved 27 May, 2019

- Entomology, science: britannica.com, Retrieved 22 January, 2019

- Entomology: newworldencyclopedia.org, Retrieved 7 April, 2019

- Taxonomic-principles: discoverlife.org, Retrieved 17 August, 2019

Chapter 2

External Anatomy of Insects

The external anatomy of insects comprises of three main segments, namely, head, thorax and abdomen. The external body of insects also comprises of three pairs of jointed legs, antennae and compound eyes. The head of insects support a pair of sensory antennae, compound eyes and appendages. The topics elaborated in this chapter will help in gaining a better perspective about these different parts belonging to the external anatomy of insects.

The Cuticle

The cuticle is a key contributor to the success of the Insecta. This inert layer provides the strong exoskeleton of body and limbs, the apodemes (internal supports and muscle attachments), and wings, and acts as a barrier between living tissues and the environment. Internally, cuticle lines the tracheal tubes, some gland ducts and the foregut and midgut of the digestive tract. Cuticle may range from rigid and armor-like, as in most adult beetles, to thin and flexible, as in many larvae. Restriction of water loss is a critical function of cuticle vital to the success of insects on land.

The cuticle is thin but its structure is complex and still the subject of some controversy. A single layer of cells, the epidermis, lies beneath and secretes the cuticle, which consists of a thicker procuticle overlaid with thin epicuticle. The epidermis and cuticle together form an integument — the outer covering of the living tissues of an insect.

The epicuticle ranges from 3 μm down to 0.1 μm in thickness, and usually consists of three layers: an inner epicuticle, an outer epicuticle, and a superficial layer. The superficial layer (probably a glycoprotein) in many insects is covered by a lipid or wax layer, sometimes called a free-wax layer, with a variably discrete cement layer external to this. The chemistry of the epicuticle and its outer layers is vital in preventing dehydration, a function derived from water-repelling (hydrophobic) lipids, especially hydrocarbons. These compounds include free and protein-bound lipids, and the outermost waxy coatings give a bloom to the external surface of some insects. Other cuticular patterns, such as light reflectivity, are produced by various kinds of epicuticular surface microsculpturing, such as close-packed, regular or irregular tubercles, ridges, or tiny hairs. Lipid composition can vary and waxiness can increase seasonally or under dry conditions. Besides being water retentive, surface waxes may deter predation, provide patterns for mimicry or camouflage, repel excess rainwater, reflect solar and ultraviolet radiation, or give species-specific olfactory cues.

The epicuticle is inextensible and unsupportive. Instead, support is given by the underlying chitinous cuticle known as procuticle when it is first secreted. This differentiates into a thicker endocuticle covered by a thinner exocuticle, due to sclerotization of the latter. The procuticle is from 10 μm to 0.5 mm thick and consists primarily of chitin complexed with protein. This contrasts with the overlying epicuticle which lacks chitin.

Chitin is found as a supporting element in fungal cell walls and arthropod exoskeletons, and is especially important in insect extracellular structures. It is an unbranched polymer of high molecular weight — an amino-sugar polysaccharide predominantly composed of β-(1-4)-linked units of N-acetyl-d-glucosamine.

Chitin molecules are grouped into bundles and assembled into flexible microfibrils that are embedded in, and intimately linked to, a protein matrix, giving great tensile strength. The commonest arrangement of chitin microfibrils is in a sheet, in which the microfibrils are in parallel. In the exocuticle, each successive sheet lies in the same plane but may be orientated at a slight angle relative to the previous sheet, such that a thickness of many sheets produces a helicoid arrangement, which in sectioned cuticle appears as alternating light and dark bands (lamellae). Thus the parabolic patterns and lamellar arrangement, visible so clearly in sectioned cuticle, represent an optical artifact resulting from microfibrillar orientation. In the endocuticle, alternate stacked or helicoid arrangements of microfibrillar sheets may occur, often giving rise to thicker lamellae than in the exocuticle. Different arrangements may be laid down during darkness com- pared with daylight, allowing precise age determination in many adult insects.

Much of the strength of cuticle comes from extensive hydrogen bonding of adjacent chitin chains. Additional stiffening comes from sclerotization, an irreversible process that darkens the exocuticle and results in the proteins becoming water-insoluble. Sclerotization may result from linkages of adjacent protein chains by phenolic bridges (quinone tanning), or from controlled dehydration of the chains, or both. Only exocuticle becomes sclerotized. The deposition of pigment in the cuticle, including deposition of melanin, may be associated with quinones, but is additional to sclerotization and not necessarily associated with it.

In contrast to the solid cuticle typical of sclerites and mouthparts such as mandibles, softer, plastic, highly flexible or truly elastic cuticles occur in insects in varying locations and proportions. Where elastic or spring- like movement occurs, such as in wing ligaments or for the jump of a flea, resilin — a "rubber-like" protein — is present. The coiled polypeptide chains of this protein function as a mechanical spring under tension or compression, or in bending.

In soft-bodied larvae and in the membranes between segments, the cuticle must be tough, but also flexible and capable of extension. This "soft" cuticle, sometimes termed arthrodial membrane, is evident in gravid females, for example in the ovipositing migratory locust, Locusta migratoria (Orthoptera: Acrididae), in which intersegmental membranes may be expanded up to 20-fold for oviposition. Similarly, the gross abdominal dilation of gravid queen bees, termites, and ants is possible through expansion of the unsclerotized cuticle. In these insects, the overlying unstretchable epicuticle expands by unfolding from an originally highly folded state, and some new epicuticle is formed. An extreme example of the distensibility of arthrodial membrane is seen in honeypot ants. In Rhodnius nymphs (Hemiptera: Reduviidae), changes in molecular structure of the cuticle allow actual stretching of the abdominal membrane to occur in response to intake of a large fluid volume during feeding.

Cuticular structural components, waxes, cements, pheromones, and defensive and other compounds are products of the epidermis, which is a near- continuous, single-celled layer beneath the cuticle.

Many of these compounds are secreted to the outside of the insect epicuticle. Numerous fine pore canals traverse the procuticle and then branch into numerous finer wax canals (containing wax filaments) within the epicuticle; this system transports lipids (waxes) from the epidermis to the epicuticular surface. The wax canals may also have a structural role within the epicuticle. Dermal glands, part of the epidermis, produce cement and wax, which is transported via larger ducts to the cuticular surface. Wax-secreting glands are particularly well developed in mealybugs and other scale insects. The epidermis is closely associated with molting — the events and processes leading up to and including ecdysis (eclosion), i.e. the shedding of the old cuticle.

Insects are well endowed with cuticular extensions, varying from fine and hair-like to robust and spine-like. Four basic types of protuberance, all with sclerotized cuticle, can be recognized on morphological, functional, and developmental grounds:

1. Spines are multicellular with undifferentiated epidermal cells;

2. Setae, also called hairs, macrotrichia, or trichoid sensilla, are multicellular with specialized cells;

3. Acanthae are unicellular in origin;

4. Microtrichia are subcellular, with several to many extensions per cell.

Setae sense much of the insect's tactile environment. Large setae may be called bristles or chaetae, with the most modified being scales, the flattened setae found on butterflies and moths (Lepidoptera) and sporadically elsewhere. Three separate cells form each seta, one for hair formation (trichogen cell), one for socket formation (tormogen cell), and one sensory cell.

There is no such cellular differentiation in multicellular spines, unicellular acanthae, and subcellular microtrichia. The functions of these types of protuberances are diverse and sometimes debatable, but their sensory function appears limited. The production of pattern, including color, may be significant for some of the microscopic projections. Spines are immovable, but if they are articulated, then they are called spurs. Both spines and spurs may bear unicellular or subcellular processes.

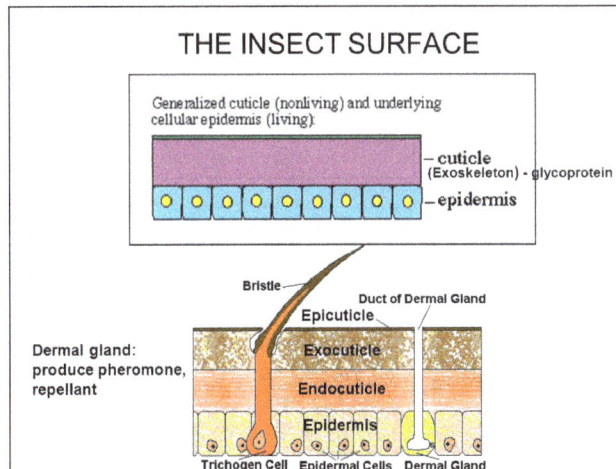

The general structure of insect cuticle.

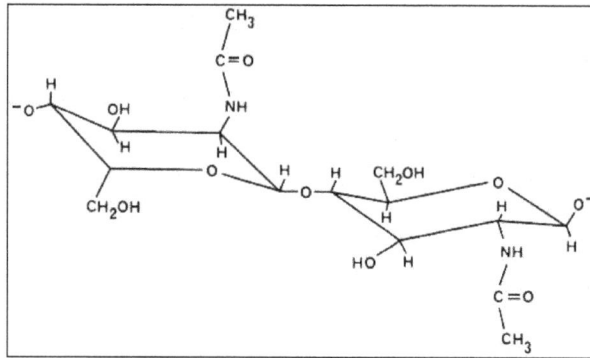

Structure of part of a chitin chain, showing two
linked units of N-acetyl-D-glucosamine.

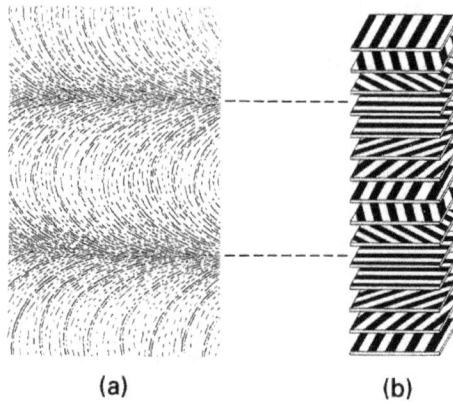

(a)   (b)

The ultrastructure of cuticle (from a transmission electron micrograph).

The arrangement of chitin microfibrils in a helicoidal array produces characteristic (though artifactual) parabolic patterns. (b) Diagram of how the rotation of microfibrils produces a lamellar effect owing to microfibrils being either aligned or non-aligned to the plane of sectioning.

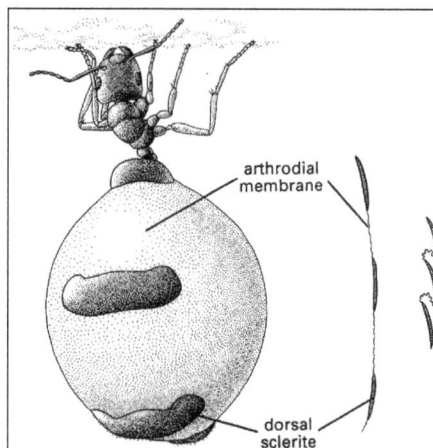

A specialized worker, or replete, of the honeypot ant, Camponotus inflatus
(Hymenoptera: Formicidae), which holds honey in its distensible abdomen
and acts as a food store for the colony.

The arthrodial membrane between tergal plates is depicted to the right in its unfolded and folded conditions.

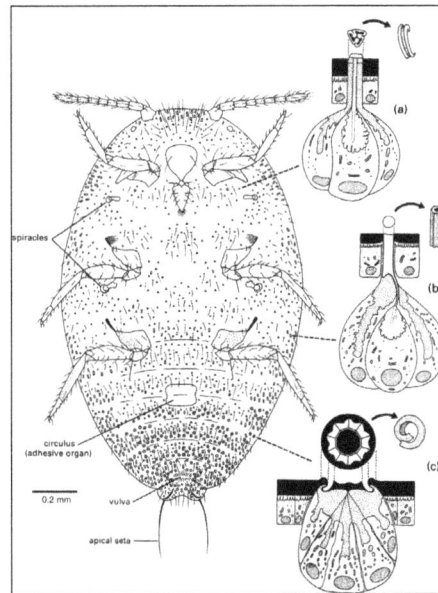

The cuticular pores and ducts on the venter of an adult female of the
citrus mealybug, Planococcus citri (Hemiptera: Pseudococcidae).

Enlargements depict the ultrastructure of the wax glands and the various wax secretions (arrowed) associated with three types of cuticular structure: (a) a trilocular pore; (b) a tubular duct; and (c) a multilocular pore. Curled filaments of wax from the trilocular pores form a protective body-covering and prevent contamination with their own sugary excreta, or honeydew; long, hollow, and shorter curled filaments from the tubular ducts and multilocular pores, respectively, form the ovisac.

The Insect Head

The insect's head is sometimes referred to as the head-capsule, and is the insect's feeding and sensory centre. It supports the eyes, antennae and and jaws of the insect. (Note -: insects do not breath through their mouths but through their thoracic and abdominal spiracles) The upper-mid portion of an insects face is called the 'frons' below this is the 'clypeus' and below this the 'labrum' to either side of which may be seen the edges of the 'mandibles' in some insects various aspects of the 'maxilliary' palps may extend beyond and or below these even when viewed from front on.

The 'frons' - that area of the face below the top two 'ocelli' and above the 'frontoclypeal sulcus' (if and when this is visible) and in between the two 'frontogenal sulci', it supports the 'pharyngeal dilator' muscles and in immature forms it bears the lower two arms of the ecdysial cleavage lines.

The 'clypeus' - that area of the face immediately below the frons (with which it may be fused in the absence of the frontoclypeal sulcus) and the frontoclypeal sulcus. It supports the 'cibarial dilator' muscles and may be divided horizontally into a 'post.' and 'anteclypeus'.

The 'labrum' - is equivalent to the insect's upper lip and is generally moveable, it articulates with the clypeus by means of the 'clypeolabral suture'.

The rest of the front of the head: that bit which is above the frons is known as the 'vertex'; the sides of the head are known as the 'gena'.

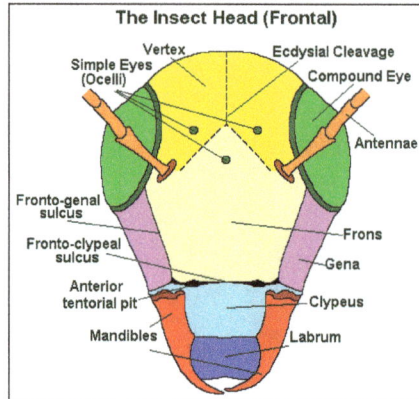

The Insect Head (Frontal)

The Antennae

The antennae are an insects primary, non-visual, sense organs, though in a few rare instances they have become adapted for other purposes such as seizing prey items (i.e. the larva of Chaoborus sp. {Diptera}) or holding females during mating (i.e. the males of Meloe sp. {Coleoptera})

Not all insects possess antennae, they are absent from the Protura.

In most insects the antennae possesses a mechanosensory organ on the pedicel (the second antennal segment) called 'Johnston's organ' and, normally, only the basal antennal segment contains intrinsic muscles. However in two orders (Diplura and Collembola) the antennae lack a 'Johnston's organ' and all but the last segment contains intrinsic muscles, thus allowing far greater controlled movement of the antennae as is demonstrated by the rolling and unrolling of the antennae observed in the Collembola Tomocerus longicornus

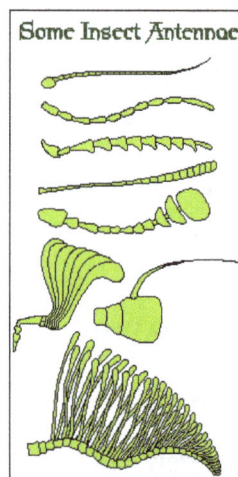

Some Insect Antennae

Antennae come in a wide variety of shapes and sizes, generally the first segment is known as the 'scape', second segment as the 'pedicel' and the rest as the flagellum. It is quite usual that the males of a species have more elaborate antennae than the females, this is because it is normally the males who have to find the females. The greater the surface area of the antennae the more dilute scents they can detect, thus male insects with feathery antennae, such as those seen in many moths, are far more sensitive than the purely filamentous ones of crickets and cockroaches.

The Eyes

Though some species of insects have been shown to be able respond to light stimulus through their cuticle, most light sensitivity occurs through one or more eyes. Insects possess two different sorts of eyes, the usually large and obviously visible compound eyes, and two varieties of ocelli or simple eyes.

Compound Eyes

Compound eyes are so named because the cornea is composed of a number of individual facets or lenses (called ommatidia), rather than a single lens as in ocelli (or our own eyes). The number of separate visual elements or ommatidia varies greatly between species as well as between the larger taxa, so that while worker ants of different species may have between 1 (Ponera punctatissima) and 600 ommatidia per single eye, adult male Odonata may have more than 28,000 per single eye.

This creates a considerable difference in the presentation of light stimulus to the insect brain, however the ability of insects to navigate the world by means of visual stimuli suggest that they have overcome the problems inherent in this multi-faceted perception. The physical differences between single-lens and multi-lens perception are excellently shown at the Bee Keepers Home Page in the section on "the world through the eye of a bee".

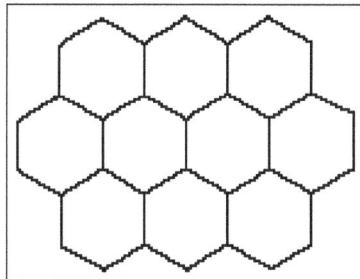

Much like our eyes, the eyes of insects, can be divided into four basic parts: the supportative material that keeps all the parts together; a light gathering part (the lens and the auxilary lens called a 'crystalline cone'); a light receptor that converts the recieved light into electrical energy; and the nerves that carry the electrical impulses to the brain for analysis. In the compound eyes of insects these parts are repeated numerous times side by side in a space saving hexagonal pattern.

The lens is formed by a transparent and colourless cuticle and it is usually biconvex. Beneath this is the crystalline cone (which is comprised of four cells called 'Semper cells' after the man who first described them). Normally this functions as a secondary lens.

The receptive parts of an insect's eye are the 'retinula cells'. Each ommatidium normally has eight retinula cells arranged to leave a central core space in the centre of the ommatidium, into which each retinula cell projects a series of microvilli (like very small fingers). These microvilli are the actual light detecting part of the cells and are collectively referred to as the rhabdomere (think cornea). The eight (or occasionally 7 or 9) rhabdomeres (sets of microvilli) form a rhabdom.

The corneal lens is supported by 'primary pigment cells' and the retinula cells and associated rhabdoms are supported by 'secondary pigment cells'. The retinula cells are connected to axons at the base of the eye, it is these which carry the information collected by the lenses and converted into electrical impulses by the rhabdom to the brain, thus allowing the insect to see.

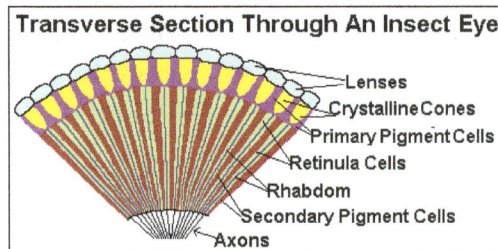

Transverse Section Through An Insect Eye
Lenses
Crystalline Cones
Primary Pigment Cells
Retinula Cells
Rhabdom
Secondary Pigment Cells
Axons

Simple Eyes

Ocelli are present in most insects to some degree, though as with all aspects of insect anatomy there is a great deal of variety in form and even in relative function. Generally they consist of five separate parts the 'cornea', the 'corneagen layer', the 'retina', the 'pigment cells', and the 'central nervous connections'.

1. The CornealLens this is a thickened area of generally transparent cuticle to the outside of the ocellus which serves as a lens.

2. The Corneagen Layer this is a single layer of specialised transparent and colourless epidermal cells which secrete the cornea.

3. The Retina this is a group of primary sensory cells which convert light into an electrical stimulus and transfer it to the; the cells are called 'retinula' cells and they are arranged in circular groups with each member of the group contributing to its portion rhabdomere to the group rhabdom. The rhabdom is the light sensitive pigment, or the part of the ocellus that converts the light into an electrical stimulus.

4. The Axon, which is the nerve link to the 'protocerebrum' and hence to the 'Corpora pedunculata' (the brain) which in turn allows the insect to use the information the ocellus produces.

5. The Pigment Cells this is a group of highly pigmented (coloured) cells variably distributed around the ocellus whose main roll would appear to be the exclusion of light from parts of the ocellus other than the cornea.

The function of the corneal lens is obscure, although it does project an image into the ocellus this image forms below the level of the light-sensitive cells, or rhabdom. Therefore the ocellus can generate no image information, however it is very sensitive to low levels of light and to changes in light intensity and scientists believe that the ocelli are useful in allowing the insect to detect the horizon, to respond quickly to changes in light intensity.

Two different forms of ocelli have been described for insects, Dorsal ocelli and Lateral ocelli.

Dorsal ocelli occur mostly in adult insects and are situated on the front of the insects face in the area of the 'frons' and or the 'epicranium', lateral ocelli generally occur on the sides of the insect head and are the form of eye most common in larval forms; there are a number of concrete differences between the two forms which can be be found explained in any competent entomological text book such as Imm's 1984.

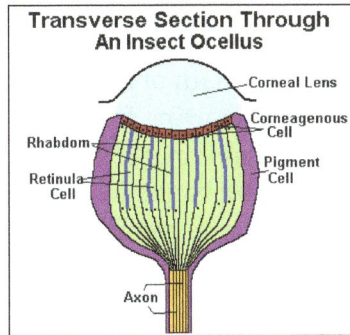

Transverse Section Through An Insect Ocellus

Corneal Lens
Corneagenous Cell
Rhabdom
Pigment Cell
Retinula Cell
Axon

Smell

Most insects communicate using smell or chemoreception and it is not surprising that they have evolved a large variety of ways of detecting the molecules involved. Insects do not have noses like us which concentrate all our sense of smell in one place, instead they have a lot of small sensory bodies scattered over their body, though they tend to have a concentration of them on their antennae. We can recognise several different common forms of chemoreceptor though these are not the only forms they can take by any means,

1. Sensilla trichoidea, hair-like structures commonly found on the feet of flies and the antennae of many insects, they are the most common form of chemoreceptor found.

2. Sensilla basiconica and Sensilla styloconica, these are peg-like or cone-like and are thicker and more solid than trichoid sensilla, these are commonly found on the antennae, though they also occur on the maxillary palps of Lepidopteran larvae and the ovipositor of the Blowfly Phormia regina.

3. Sensilla coeloconica or pit-peg organs these are always situated in a pit as their common name suggests, unlike the previous two which project above the insects cuticle. They are common on a variety of insect antennae and and in Apis mellifera (the Honey Bee) they detect Carbon Dioxide.

4. Sensilla placodea these differ from the first three in that they consist of a flat plate of cuticle, they occur on the antennae of various Aphids and Apis mellifera (the Honey Bee).

Mouthparts

Labrum   Mandibles   Maxillae   Labium

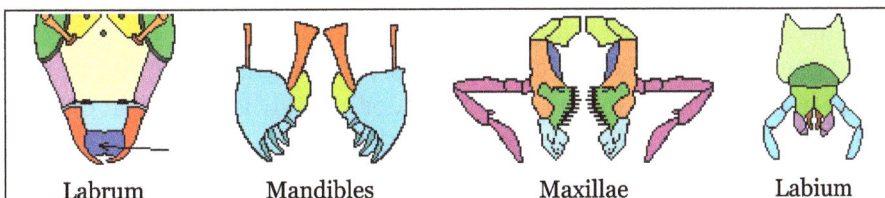

In the primitive form the insect mouth articulates (moves) from side to side in a horizontal plane, rather than vertically as do ours. In those groups of insects that evolved later the basic mouthparts shown below have become highly modified, in the butterflies and moths they are transformed into a long flexible straw, in the blood and sap sucking insects of various orders they have become a hypodermic needle and in most of the flies you see from day to day they evolved into a extensible and highly effective sponge. However in all these cases, all, or most of the basic parts are still there, see the tranverse section of the needle like mouth of the mosquito right as an example. The Insect mouth consists of a number of parts which, starting from the foremost, are called 1) The Labrum, 2) The Mandibles, 3) The maxillae, 4) The Labium, 5) The Hypopharynx or tongue. The Labrum is normally a simply structure that is equivalent to the insect's upper lip and is generally moveable, it articulates with the clypeus by means of the 'clypeolabral suture'.

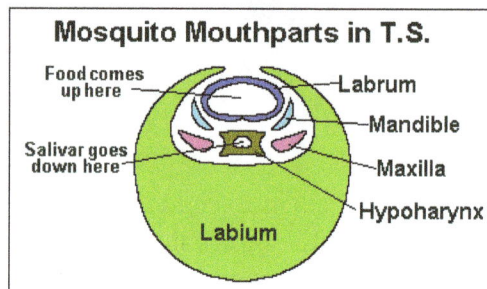

Mosquito Mouthparts in T.S.

Mandibles

The mandibles, with the maxillae, the labial palps and in some species the hypopharynx constitute the moveable aspects of the insect mouth, the mandibles and the maxillae are the equivalent of jaws with the exception that they move transversely (from side to side).

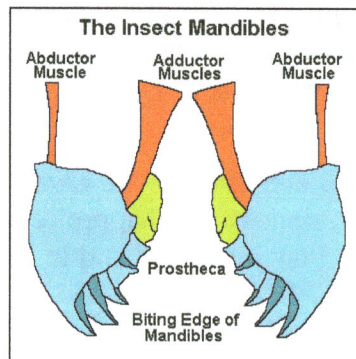

The Insect Mandibles

The mandibles show great variety within the insect orders and like our more familiar teeth they are hard and show variation in accordance with diet, thus they are sharp edged in carnivores, being extremely sickle like in the ant Aceton burchelli, whilst being adapted for crushing and chewing in herbivores. They have become a secondary sexual characteristic and are extremely large in some Beetles (i.e. in the genus Chiasognathus and Lucanus [Stag Beetles]). In other orders they have become residual as in the adults of some Lepidoptera, or entirely absent as in the adults of the rest of the Lepidoptera, Trichoptera, Ephemeroptera and the Diptera.

Mandibles are are used not only for feeding but also for attack and defence, becoming extremely exaggerated in various species of termites ants (i.e. Atta texanus), and for manipulation of materials as in the nest building insects particularly the Hymenopteran Bees, Ants and Wasps.

Maxillae

The maxillae are a pair of modified limbs which work behind the mandibles and in front of the labium as a pair of accessory jaws. They are composed of the following parts.

1. The Cardo, this is the piece nearest the head capsule and in some species of insect it is the only part of the maxillae that is connected to the head.

2. The Stypes this is central bulk of the maxillae and supports the;

3. Palpifer, which in turn supports the;

4. Maxilliary palp, which has one to seven segments and is mainly used as a sensory organ.

5. The Lacinia is situated at the distal end of the Stypes, is often serrated or toothed and serves to aid the eating process both by holding and masticating the food. It is boarded distally (to the far side in relation to the overall body) by;

6. The Galea and proximally (to the near-side in relationship to the body).

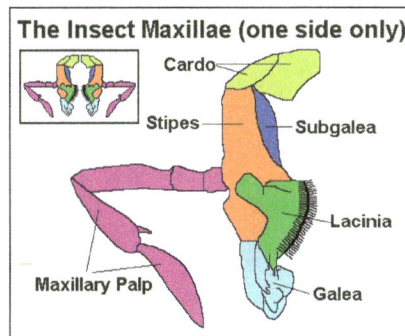

The Insect Maxillae (one side only)

Labium

The Labium results from the fusion of a pair of limbs and serves a purpose similar to our lower lip for the insects. The main body of the labium is divided into three parts the central 'mentum' which is boarded on either side by the 'submentum' proximally which hinges with the head, and the prementum distally. The prementum supports two pairs of lobes known as the 'glossae' and to the outside of them the 'paraglossae', and a pair of labial palps which are primarily sensory in function. The glossae and paraglossae may be fused, with one or the other considerably reduced, in which case the whole thing is known as the 'ligula'.

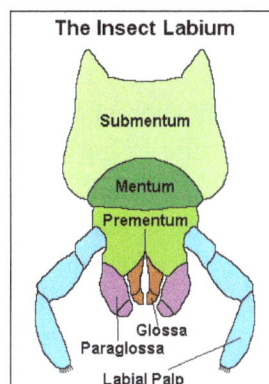

The Insect Labium

The hyper-pharynx or tongue is found behind the mouth, and has the salivary ducts at its base, in most Diptera (true flies) and Hemiptera (true bugs) it has become highly modified and serves as the main feeding organ, in many cases combining with the rest of the mouthparts to form a stylette or piercing organ.

Mouthparts

The mouthparts are formed from appendages of all head segments except the second. In omnivorous insects, such as cockroaches, crickets, and earwigs, the mouthparts are of a biting and chewing type (mandibulate) and resemble the probable basic design of ancestral pterygote insects more closely than the mouthparts of the majority of modern insects. Extreme modifications of basic mouthpart structure, correlated with feeding specializations, occur in most Lepidoptera, Diptera, Hymenoptera, Hemiptera, and a number of the smaller orders. Here we first discuss basic mandibulate mouthparts, as exemplified by the European earwig, Forficula auricularia (Dermaptera: Forficulidae), and then describe some of the more common modifications associated with more specialized diets.

There are five basic components of the mouthparts:

1. Labrum, or "upper lip", with a ventral surface called the epipharynx;

2. Hypopharynx, a tongue-like structure;

3. Mandibles, or jaws;

4. Maxillae (singular: maxilla);

5. Labium, or "lower lip".

The labrum forms the roof of the preoral cavity and mouth and covers the base of the mandibles; it may be formed from fusion of parts of a pair of ancestral appendages. Projecting forwards from the back of the preoral cavity is the hypopharynx, a lobe of probable composite origin; in apterygotes, earwigs, and nymphal mayflies the hypopharynx bears a pair of lateral lobes, the superlinguae (singular: superlingua). It divides the cavity into a dorsal food pouch, or cibarium, and a ventral salivarium into which the salivary duct opens. The mandibles, maxillae, and labium are the paired appendages of segments 4–6 and are highly variable in structure among insect orders; their serial homology with walking legs is more apparent than for the labrum and hypopharynx.

The mandibles cut and crush food and may be used for defense; generally they have an apical cutting edge and the more basal molar area grinds the food. They can be extremely hard (approximately 3 on Moh's scale of mineral hardness, or an indentation hardness of about 30 kg mm-2) and thus many termites and beetles have no physical difficulty in boring through foils made from such common metals as copper, lead, tin, and zinc. Behind the mandibles lie the maxillae, each consisting of a basal part composed of the proximal cardo and the more distal stipes and, attached to the stipes, two lobes — the mesal lacinia and the lateral galea — and a lateral, segmented maxillary palp, or palpus (plural: palps or palpi). Functionally, the maxillae assist the mandibles in processing food; the pointed and sclerotized lacinae hold and macerate the food, whereas the galeae and palps bear sensory setae (mechanoreceptors) and chemoreceptors which sample items before ingestion. The appendages of the sixth segment of the head are fused with the sternum to form the labium, which is believed to be homologous to the second maxillae of Crustacea. In prognathous insects, such as the earwig, the labium attaches to the ventral surface of the head via a ventromedial sclerotized plate called the gula. There are two main parts to the labium: the proximal postmentum, closely connected to the posteroventral surface of the head and sometimes subdivided into a submentum and mentum; and the free distal prementum, typically bearing a pair of labial palps lateral to two pairs of lobes, the mesal glossae (singular: glossa) and the more lateral paraglossae (singular: paraglossa). The glossae and paraglossae, including sometimes the distal part of the prementum to which they attach, are known collectively as the ligula; the lobes may be variously fused or reduced as in Forficula, in which the glossae are absent. The prementum with its lobes forms the floor of the preoral cavity (functionally a "lower lip"), whereas the labial palps have a sensory function, similar to that of the maxillary palps.

During insect evolution, an array of different mouthpart types have been derived from the basic design described above. Often feeding structures are characteristic of all members of a genus, family, or order of insects, so that knowledge of mouthparts is useful for both taxonomic classification and identification, and for ecological generalization. Mouthpart structure is categorized generally according to feeding method, but mandibles and other components may function in defensive combat or even male—male sexual contests, as in the enlarged mandibles on certain male beetles (Lucanidae). Insect mouthparts have diversified in different orders, with feeding methods that include lapping, suctorial feeding, biting, or piercing combined with sucking, and filter feeding, in addition to the basic chewing mode.

The mouthparts of bees are of a chewing and lapping type. Lapping is a mode of feeding in which liquid or semi-liquid food adhering to a protrusible organ, or "tongue", is transferred from substrate to mouth. In the honey bee, Apis mellifera (Hymenoptera: Apidae), the elongate and fused labial glossae form a hairy tongue, which is surrounded by the maxillary galeae and the labial palps to form a tubular proboscis containing a food canal. In feeding, the tongue is dipped into the nectar or honey, which adheres to the hairs, and then is retracted so that adhering liquid is carried into the space between the galeae and labial palps. This back-and-forth glossal movement occurs repeatedly. Movement of liquid to the mouth apparently results from the action of the cibarial pump, facilitated by each retraction of the tongue pushing liquid up the food canal. The maxillary laciniae and palps are rudimentary and the paraglossae embrace the base of the tongue, directing saliva from the dorsal salivary orifice around into a ventral channel from whence it is transported to the flabellum, a small lobe at the glossal tip; saliva may dissolve solid or semi-solid sugar. The sclerotized, spoon-shaped mandibles lie at the base of the proboscis and have a variety of functions, including the manipulation of wax and plant resins for nest construction, the feeding of larvae and the queen, grooming, fighting, and the removal of nest debris including dead bees.

Most adult Lepidoptera and some adult flies obtain their food solely by sucking up liquids using suctorial (haustellate) mouthparts that form a proboscis or rostrum (Box 15.5). Pumping of the liquid food is achieved by muscles of the cibarium and/or pharynx. The proboscis of moths and butterflies, formed from the greatly elongated maxillary galeae, is extended by increases in hemolymph ("blood") pressure. It is loosely coiled by the inherent elasticity of the cuticle, but tight coiling requires contraction of intrinsic muscles. A cross-section of the proboscis shows how the food canal, which opens basally into the cibarial pump, is formed by apposition and interlocking of the two galeae. The proboscis of some male hawk- moths (Sphingidae), such as that of Xanthopan morgani, can attain great length.

A few moths and many flies combine sucking with piercing or biting. For example, moths that pierce fruit and exceptionally suck blood (species of Noctuidae) have spines and hooks at the tip of their proboscis which are rasped against the skins of either ungulate mammals or fruit. For at least some moths, penetration is effected by the alternate protraction and retraction of the two galeae that slide along each other. Bloodfeeding flies have a variety of skin-penetration and feeding mechanisms. In the "lower" flies such as mosquitoes and black flies, and the Tabanidae (horse flies, Brachycera), the labium of the adult fly forms a non-piercing sheath for the other mouthparts, which together contribute to the piercing structure. In contrast, the biting calyptrate dipterans (Brachycera: Calyptratae, e.g. stable flies and tsetse flies) lack mandibles and maxillae and the principal piercing organ is the highly modified labium. Mouthparts of adult Diptera are described in Box 15.5.

Other mouthpart modifications for piercing and sucking are seen in the true bugs (Hemiptera), thrips (Thysanoptera), fleas (Siphonaptera), and sucking lice (Phthiraptera: Anoplura). In each order different mouthpart components form needle-like stylets capable of piercing the plant or animal tissues upon which the insect feeds. Bugs have extremely long, thin paired mandibular and maxillary stylets, which fit together to form a flexible stylet-bundle containing a food canal and a salivary canal (Box 11.8). Thrips have three stylets — paired maxillary stylets (laciniae) plus the left mandibular one. Sucking lice have three stylets — the hypopharyngeal (dorsal), the salivary (median), and the labial (ventral) — lying in a ventral sac of the head and opening at a small eversible proboscis armed with internal teeth that grip the host during blood-feeding. Fleas possess a single stylet derived from the epipharynx, and the laciniae of the maxillae form two long cutting blades that are ensheathed by the labial palps. The Hemiptera and the Thysanoptera are sister groups and belong to the same assemblage as the Phthiraptera, but the lice at least had a psocopteroid-like ancestor, presumably with mouthparts of a more generalized, mandibulate type. The Siphonaptera are distant relatives of the other three taxa; thus similarities in mouthpart structure among these orders result largely from parallel or, in the case of fleas, convergent evolution.

Slightly different piercing mouthparts are found in antlions and the predatory larvae of other lacewings (Neuroptera). The stylet-like mandible and maxilla on each side of the head fit together to form a sucking tube, and in some families (Chrysopidae, Myrmeleontidae, and Osmylidae) there is also a narrow poison channel. Generally, labial palps are present, maxillary palps are absent, and the labrum is reduced. Prey is seized by the pointed mandibles and maxillae, which are inserted into the victim; its body contents are digested extra-orally and sucked up by pumping of the cibarium.

A unique modification of the labium for prey capture occurs in nymphal damselflies and dragonflies (Odonata These predators catch other aquatic organisms by extending their folded labium (or "mask") rapidly and seizing mobile prey using prehensile apical hooks on modified labial palps. The labium is hinged between the prementum and postmentum and, when folded, covers most of the underside of the head. Labial extension involves the sudden release of energy, produced by increases in blood pressure brought about by the contraction of thoracic and abdominal muscles, and stored elastically in a cuticular click mechanism at the prementum—postmentum joint. As the click mechanism is disengaged, the elevated hydraulic pressure shoots the labium rapidly forwards. Labial retraction then brings the captured prey to the other mouthparts for maceration.

Filter feeding in aquatic insects has been studied best in larval mosquitoes (Diptera: Culicidae), black flies (Diptera: Simuliidae), and net-spinning caddisflies (Trichoptera: many Hydropsychoidea and Philopotamoidea), which obtain their food by filtering particles (including bacteria, microscopic algae, and detritus) from the water in which they live. The mouthparts of the dipteran larvae have an array of setal "brushes" and "fans", which generate feeding currents or trap particulate matter and then move it to the mouth. In contrast, the caddisflies spin silk nets that filter particulate matter from flowing water and then use their mouthpart brushes to remove particles from the nets. Thus insect mouthparts are modified for filter feeding chiefly by the elaboration of setae. In mosquito larvae the lateral palatal brushes on the labrum generate the feeding currents; they beat actively, causing particle-rich surface water to flow towards the mouth- parts, where setae on the mandibles and maxillae help to move particles into the pharynx, where food boluses form at intervals.

In some adult insects, such as mayflies (Ephemeroptera), some Diptera (warble flies), a few moths (Lepidoptera), and male scale insects (Hemiptera: Coccoidea), mouthparts are greatly reduced and non- functional. Atrophied mouthparts correlate with short adult lifespan.

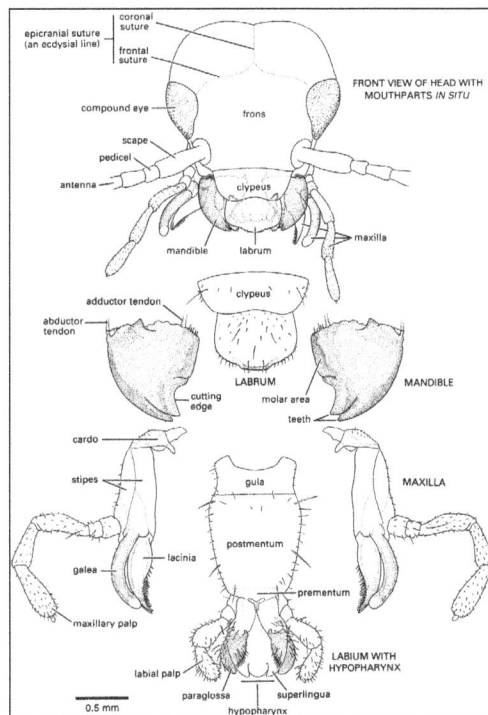

Frontal view of the head and dissected mouthparts of an adult of the
European earwig, Forficula auricularia (Dermaptera: Forficulidae).

Note that the head is prognathous and thus a gular plate, or gula, occurs in the ventral neck region.

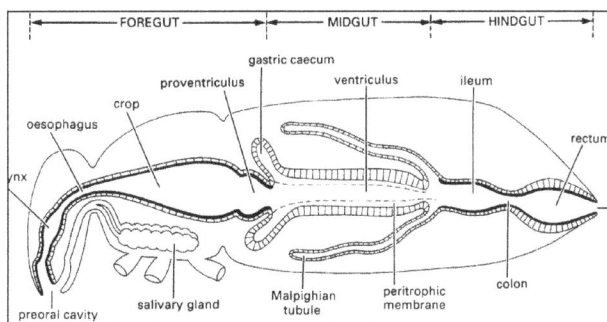

Preoral and anterior foregut morphology in (a) a generalized orthopteroid insect
and (b) a xylem-feeding cicada.

Musculature of the mouthparts and the (a) pharyngeal or (b) cibarial pump are indicated but not fully labeled. Contraction of the respective dilator muscles causes dilation of the pharynx or cibarium and fluid is drawn into the pump chamber. Relaxation of these muscles results in elastic return of the pharynx or cibarial walls and expels food upwards into the oesophagus.

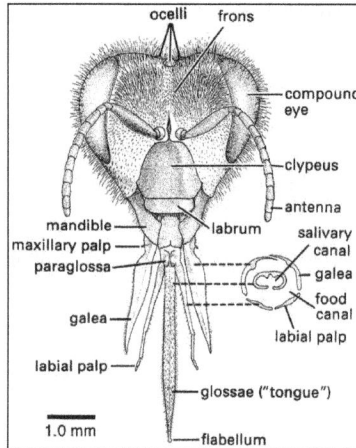

Frontal view of the head of a worker honey bee, Apis mellifera (Hymenoptera: Apidae), with transverse section of proboscis showing how the "tongue" (fused labial glossae) is enclosed within the sucking tube formed from the maxillary galeae and labial palps.

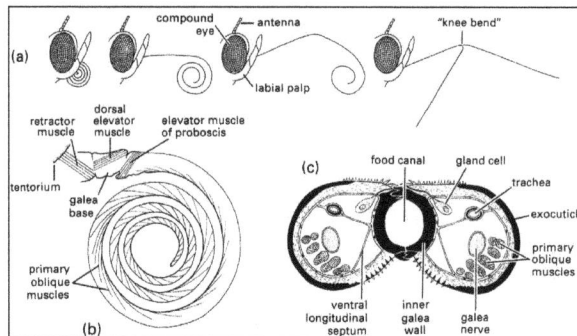

Mouthparts of the cabbage white or cabbage butterfly, Pieris rapae (Lepidoptera: Pieridae).

(a) Positions of the proboscis showing, from left to right, at rest, with proximal region uncoiling, with distal region uncoiling, and fully extended with tip in two of many possible different positions due to flexing at "knee bend". (b) Lateral view of proboscis musculature. (c) Transverse section of the proboscis in the proximal region.

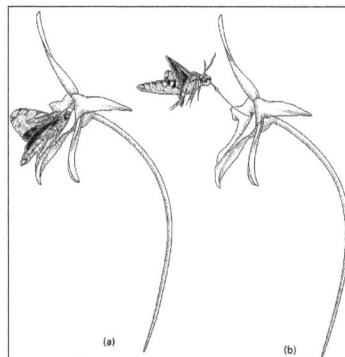

A male hawkmoth of Xanthopan morgani praedicta (Lepidoptera: Sphingidae) feeding from the long floral spur of a Malagasy star orchid.

Angraecum sesquipedale:

(a) full insertion of the moth's proboscis; (b) upward flight during withdrawal of the proboscis with the orchid pollinium attached.

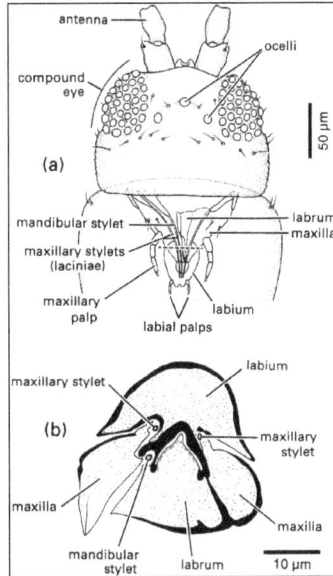

Head and mouthparts of a thrips, Thrips australis
(Thysanoptera: Thripidae).

(a) Dorsal view of head showing mouthparts through prothorax. (b) Transverse section through proboscis. The plane of the transverse section is indicated by the dashed line in (a).

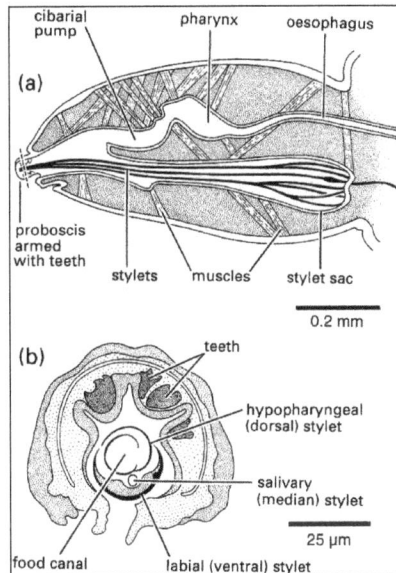

Head and mouthparts of a sucking louse, Pediculus
(Phthiraptera: Anoplura: Pediculidae).

(a) Longitudinal section of head (nervous system omitted). (b) Transverse section through eversible proboscis. The plane of the transverse section is indicated by the dashed line in (a).

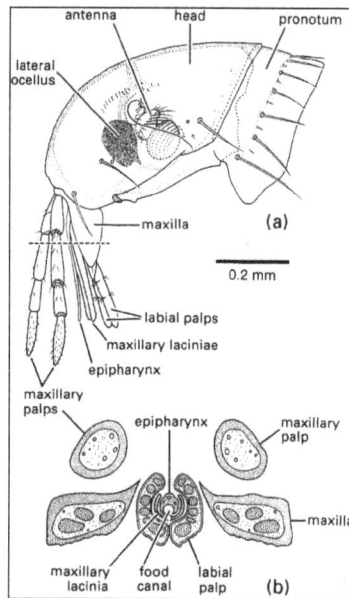

Head and mouthparts of a human flea, Pulex
irritans (Siphonaptera: Pulicidae):

(a) lateral view of head; (b) transverse section through mouthparts. The plane of the transverse
section is indicated by the dashed line in (a).

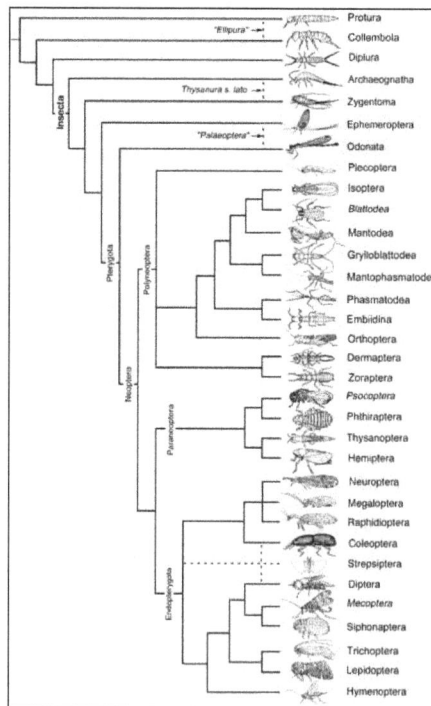

Cladogram of postulated relationships of extant hexapods, based on combined morphological and
nucleotide sequence data. Italicized names indicate paraphyletic taxa.

Broken lines indicate uncertain relationships. Thysanura sensu lato refers to Thysanura in the
broad sense.

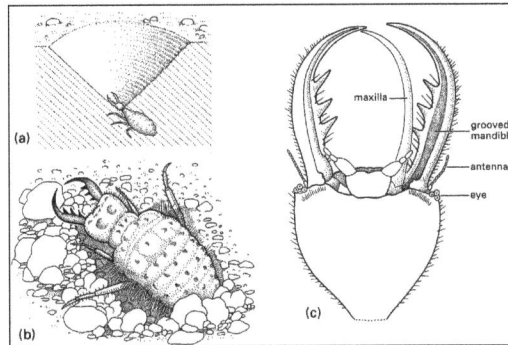

An antlion of Myrmeleon (Neuroptera: Myrmeleontidae)

(a) larva in its pit in sand; (b) detail of dorsum of larva; (c) detail of ventral view of larval head showing how the maxilla fits against the grooved mandible to form a sucking tube.

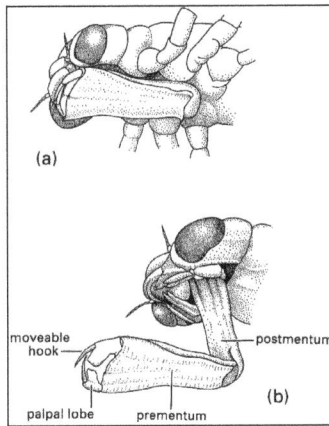

Ventrolateral view of the head of a dragonfly nymph
(Odonata: Aeshnidae: Aeshna) showing the labial "mask":

(a) in folded position, and (b) extended during prey capture with opposing hooks of the palpal lobes forming claw-like pincers.

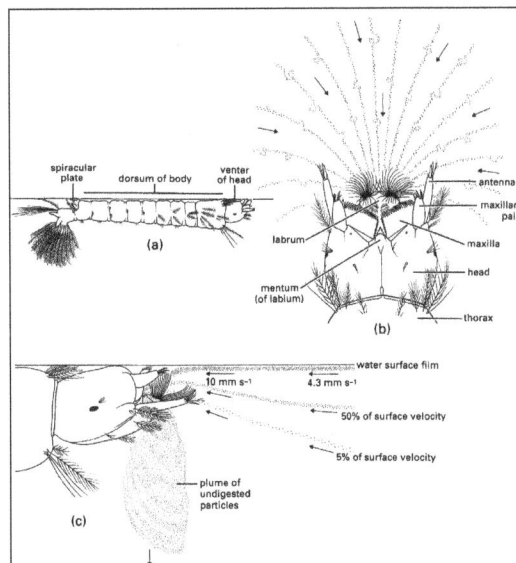

The mouthparts and feeding currents of a mosquito larva of Anopheles quadrimaculatus (Diptera: Culicidae).

(a) The larva floating just below the water surface, with head rotated through 180° relative to its body (which is dorsum-up so that the spiracular plate near the abdominal apex is in direct contact with the air). (b) Viewed from above showing the venter of the he ad and the feeding current generated by setal brushes on the labrum (direction of water movement and paths taken by surface particles are indicated by arrows and dotted lines, respectively). (c) Lateral view showing the particle-rich water being drawn into the preoral cavity between the mandibles and maxillae and its downward expulsion as the outward current.

The Insect Abdomen

The insect abdomen is built up of a series of concave upper integumental plates known as 'tergites' and convex lower integumental plates known as 'sternites', the whole being held together by a tough yet stretchable membrane. It contains the insects digestive tract and reproductive organs, it consists of eleven segments in most orders of insects though the 11th segment is absent in the adult of most the higher orders. In many of the Hymenoptera, and several other orders, the number of abdominal segments is reduced and in the Honey Bee only seven segments are visible. In the 'Collembola' (Springtails) the abdomen has only six segments. In the Hymenoptera there is a constriction where the 1st and 2nd abdominal segments meet, this is called the waste, and the remaining portion of the abdomen is called the gaster.

Unlike other Arthropods the the insects possess no legs on the abdomen in the adult form, though the 'Protura' do have rudimentary leg-like appendages on the first three abdominal segments. Many larval insects particularly the 'Lepidoptera' and the 'Symphyta' (Sawflies) have appendages called 'pseudo' or prolegs on their posterior abdominal segments as well as their more familiar thoracic legs these allow them to grip onto the edges of plant leaves as they walk around.

The Muscles

Like us Insects need muscles in order to move the various different bits of their bodies around, however insects have their muscles attached to the inside of their skeleton because like all the arthropods they have their skeletons on the outside of their body. The inside of an insect's exoskeleton has special contours and bits and bobs on it which project inwards and allow for muscles to be attached and to help give them leverage, these projections are called 'apodemes'. The musculature of even the smallest insect can be as complicated as our own and makes for a fascinating study of design in miniature. The muscles of insects are generally light grey or translucent, unlike ours which appear red. This is because insects lack both the blood system that we have and the haemoglobin that makes our blood and hence our muscles red.

The Heart and the Blood

The haemolymph (blood) of insects flows freely around the inside of their bodies. Because an insect's haemolymph is not responsible for the transmission of oxygen to its cells, this is the job of the Trachaea, and therefore does not contain haemoglobin, it is not red. Normally it is a watery green colour, though it is pigmented (coloured) in some species. This haemolymph is a sort of soup rich in nutrients that flows around the inside of the insects body allowing the various organs to get at whatever resources they need and into which they dump their waste products. These waste products are later removed from the haemolymph by the Malpighian tubules. Because insects do not have veins and arteries like us and the rest of the vertebrates, (they do not need to), they do not have a complicated heart like ours either. The insect heart it is basically a tube, sealed at one end, which runs along their back. It beats regularly thus swishing the blood in and out and around the body. In some cases it has inlets, with one way valves, and in others it has outlets as well. The Aorta is simple a tube that runs from the heart towards the brain, it is open at the end near the brain. You can get some idea how the heart of an insect works by cutting the nozzle off the top of an empty washing-up-liquid bottle and taking it into the bath with you. Hold it under the water and squeeze it a few times and watch how the water swirls around.

Digestion

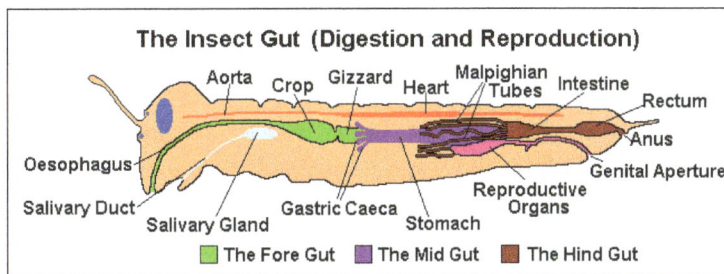

The Digestive system of an insect is usually a long straight tube running from the mouth to the anus, it is often divided into the 'fore gut', the 'mid gut' and the 'hind gut'. Immediately behind the mouth are the Salivary Glands, in most species these secrete saliva, generally a watery fluid that lubricates the food and contains a few enzymes to begin the processes of digestion. However in some carnivorous insects the saliva is composed entirely of digestive enzymes, this applies particularly to those with external digestion of the food. In other insects the salivary glands have become modified for purposes that have nothing to do with digestion. In Lepidopteran caterpillars and Caddisfly larvae they have been converted to the production of silk, while in the Queen Honey Bee they are called the mandibular glands and secrete hormones.

The Fore Gut

The fore gut is generally considered to consist of four sections, the Pharynx, the Oesophagus, the Crop and the Proventriculus. It is also known as the Stomodaeum.

The pharynx is the first part of the fore gut and apart from being a tube that connects the interior of the mouth area (sometimes known as the 'Buccal Cavity) with the more inward parts of the gut it sometimes serves as a pump to suck up the liquefied food of those insects which feed by means external digestion. The Oesophagus is basically a tube leading to the mid gut via the crop and the proventriculus or gizzard. The crop is simply a storage area and the proventriculus, or gizzard, is a muscular extension of the crop. In those insects which feed on solid foods it is used to grind the food up into smaller particles, it can also serve as a filter to keep oversized particles out of the main digestive tract and as a valve controlling the flow of food into the midgut. The fore gut and the mid gut are separated by the 'stomodeal or cardiac valve'.

Mid Gut

The midgut (called the Mesenteron in some books) runs from the 'digestive or gastric caeca', a series of stubby pointed tubes leading off from the stomach to just before the Malpighian tubules, a series of long thin tubes. In between the two of these is the stomach, or ventriculus, which is the area of most active digestion. The gastric caeca serve to increase the surface area of the midgut, thus increasing both its ability to secrete digestive enzymes and its ability to extract useful products from the partially digested food. The useful proteins, vitamins and fats that are released by the digestive processes pass across the wall of the midgut into the body cavity. The mid gut is lined by a semipermeable membrane composed of protein and chitin, like the cuticle, which allows the passage of liquids and dissolved substances to the midgut wall while preventing the passage of solid food particles, it is continually worn away by the passage of food through the gut and replaced by the epithelial cells of the mid gut wall. The mid gut and the hindgut are separated by the 'proctodeal valve'.

Hind Gut

From the mid gut food passes to the hind gut (called the Proctodaeum in some books). The hind gut comprises the 'intestines' which is where much of the diffusion into the the insects body occurs. The 'rectum' which compresses the undigested food and waste products, extracts more water from this if necessary before it is passed out through the 'anus' as faeces.

Arising from and the foremost part of the hind gut are the Malpighian tubules (named after Malpighi who discovered them) are not really to do with digestion at all but with elimination. They act like our kidneys and extract metabolic waste products (mostly nitrogenous ones such as urea, and uric acid) from the circulating body fluid called the haemolymph and excrete them into the intestines which is the first part of the hind gut.

Though insects possess a large number of digestive enzymes, they are often helped by the presence of symbiotic micro-organisms, such as protozoa in the case of the termites and some primitive cockroaches which feed on wood, and bacteria in the wax moth Galleria mellonella which feeds on the wax that honey bees Apis mellifera uses to make the combs in its hives.

Reproductive Structures

Most insect species are bisexual, i.e. there are males and females in most species, these often look very different and have even been mistaken for different species in the past, some species are capable of reproduction without males, the eggs are unfertilised but develop and hatch into nymphs or larvae that are always female themselves, this is called 'parthenogenesis'.

Externally the sexual organs, called genitalia, of a female insect generally consist of an 'ovipositor' which is often encased in a pair of filaments called a 'sheath' and is which is used to by the female to put her eggs where she wants. Its form very greatly throughout the Insecta (i.e. the whole order of insects). The ovipositor of the Diptera (True Flies) is functionally similar i.e. it is used to lay eggs, but is morphologically distinct i.e. it arises or is made from different parts of the insects anatomy and should be called a 'pseudovipositor'. The median part of the oviduct which receives the aedeagus during mating is called the 'vagina'.

Externally the sexual organs of the male, also called genitalia, consist of a pair of 'claspers' which the male uses to hold onto the females genitalia and an intromittant organ called the 'aedeagus' which is the means by which the male passes the sperm onto the female.

Internally the female reproductive organs consist of a pair of ovaries which contain the ovarioles which is where the eggs or ova are formed, the bursa copulatrix which is where the sperm is first received)in those insects which have it) and a spermatheca which is where the sperm is stored. There are also various tubes down which the ova travel on their way from the ovaries to the outside world, fertilisation occurs in the common oviduct after the the egg has received its shell or the 'chorion'. To facilitate this the shell contains a very small opening at one end called the micropyle which allows the sperm to enter. As well as tubes there are several important glands some of which (spermathecal glands) allow the female to keep the sperm alive and viable for a long time, as much as 20 years in some social insects (Ants and Bees); and some of which (collaterial glands) secrete the substances which allow the female to stick the eggs where she wants then to stay i.e. underneath a leaf, or to protect the eggs as in the ootheca produced by the Cockroaches and Mantids.

Internally the male reproductive organs consist of a pair of 'testes' containing the 'testicular follicles' where the spermatozoa are made, the 'vas deferens' which is the tube down which the sperm travels, a 'seminal vesicle' which is where the sperm is stored prior to mating, and accessory glands which supply seminal fluid for additional volume and to nourish the sperm before and during their journey.

The Insect Thorax

The skeleton of the thoracic segments is modified to give efficient support for the legs and wings, and the musculature is adapted to produce the movements of these appendages.

Morphology of the Thorax

Tergum

The tergum of the prothoracic segment is known as the pronotum. It is often small serving primarily for attachment of the muscles of the first pair of legs, but in Orthoptera, Blattodea and Coleoptera it forms a large plate affording some protection to the pterothoracic segments. The meso- and meta-nota are relatively small in wingless insects and larvae, but in winged insects they become modified for the attachment of the wings. In the majority of winged insects, the downward movement of the wings depends on an upwards distortion of the dorsal wall of the thorax. This is made possible by a modification of the basic segmental arrangement.

Various strengthening ridges develop on the tergum of a wing-bearing segment. These are local adaptations to the mechanical stresses imposed by the wings and their muscles. The ridges appear externally as sulci which divide the notum into areas. Often a transverse sulcus divides the notum into an anterior prescutum and a scutum, while a V-shaped sulcus posteriorly cuts off the scutellum. The various elements of the tergum as a result of modifications leading to subdivisions include acrotergite, prescutum, scutum, scutellum along an anterior to posterior plane. The sulci of the tergum include the antecostal sulcus separating the acrotergite and prescutum; transverse sulcus separating the prescutum from the scutum; and the scutoscutellar sulcus separating the scutum from the scutellum.

Sternum

The primary sclerotizations on the ventral side are segmental and inter segmental plates which often remain separate in the thorax. The intersegmental sclerite is produced internally into a spine and is called the spinasternum, while the segmental sclerite is called the eusternum.

The sternum of the pterothoracic segments does not differ markedly from that of the prothorax, but usually the basisternum is bigger, providing for the attachment of the large dorsoventral flight muscles. The sternum is attached to the pleuron by pre- and post-coxal bridges.

Arising from the eusternum are a pair of apophyses, the so-called sternal apophyses. The origins of these on the sternum are marked externally by pits joined by a sulcus so that the eusternum is divided into a basisternum and sternellum, while in higher insects the two apophyses arise together in the midline and only separate internally, forming a Y-shaped furca.

Distally the sternal apophyses are associated with the inner ends of the pleural apophyses, usually being connected to them by short muscles. This adds rigidity to the thorax, while variation in the degree of contraction of the muscles makes this rigidity variable and controllable. The sternal apophyses also serve for the attachment of the bulk of the ventral longitudinal muscles.

Pleuron

The pleural regions are membranous in many larval insects, but typically become sclerotized in the adult. Basically there are probably three pleural sclerites, one ventral and two dorsal, which may originally have been derived from the coxa. Above the coxa the pleuron develops a nearly vertical strengthening ridge, the pleural ridge, marked by the pleural sulcus externally. This divides the pleuron into an anterior episternum and a posterior epimeron. The pleural ridge is particularly well developed in the wing-bearing segments, where it continues dorsally into the pleural wing process which articulates with the second axillary sclerite in the wing base.

In front of the pleural process in the membrane at the base of the wing and only indistinctly separated from the episternum are one or two basalar sclerites, while in a comparable position behind the pleural process is a well-defined subalar sclerite. Muscles concerned with the movement of the wings are inserted into these sclerites.

Typically, there are two pairs of spiracles on the thorax. These are in the pleural regions and are associated with the mesothoracic and metathoracic segments. The mesothoracic spiracle often occupies a position on the posterior edge of the propleuron, while the smaller metathoracic spiracle may similarly move on to the mesothorax. Diplura have three or four pairs of thoracic spiracles.

Heterojapyx, for instance, has two pairs of mesothoracic and two pairs of metathoracic spiracles.

Muscles of the Thorax

The longitudinal muscles of the thorax, as in the abdomen, run from one antecostal ridge to the next. They are relatively poorly developed in sclerotized larvae, in adult Odonata, Blattodea and Isoptera which have direct wing depressor muscles, and also in secondarily wingless groups such as Siphonaptera. In these cases, they tend to telescope one segment into the next, while the more lateral muscles rotate the segments relative to each other. In unsclerotized insects, contraction of the longitudinal muscles shortens the segment.

In most winged insects, however, the dorsal longitudinal muscles are the main wing depressors and they are well developed, running from phragma to phragma so that their contraction distorts the segments. The ventral longitudinal muscles run mainly from one sternal apophysis to the next in adult insects, producing some ventral telescoping of the thoracic segments. Dorso-ventral muscles run from the tergum to the pleuron or sternum. They are primitively concerned with rotation or compression of the segment, but in winged insects they are important flight muscles. In larval insects an oblique intersegmental muscle runs from the sternal apophysis to the anterior edge of the following tergum or pleuron, but in adults it is usually only present between prothorax and mesothorax.

Insect Legs and Locomotion

Insects typically have three pairs of legs, one pair on each of the thoracic segments. From this, the alternative name for insects, the 'hexapods', is derived, although not all hexapods are now regarded as insects.

Basic Structure of the Legs

Each leg consists typically of six segments, articulating with each other by mono- or di-condylic articulations set in a membrane, the corium. The six basic segments are:

1. Coxa – the coxa is often in the form of a truncated cone and articulates basally with the wall of the thorax. The part of the coxa bearing the articulations is often strengthened by a ridge indicated externally by the basi-costal sulcus which marks off the basal part of the coxa as the basicoxite. The basicoxite is divided into anterior and posterior parts by a ridge strengthening the articulation, the posterior part being called the meron.

2. Trochanter – the trochanter is a small segment with a dicondylic articulation with the coxa such that it can only move in the vertical plane. In Odonata there are two trochanters and this also appears to be the case in Hymenoptera, but here the apparent second trochanter is, in fact, a part of the femur.

3. Femur – the femur is often small in larval insects, but in most adults it is the largest and stoutest part of the leg. It is often more or less fixed to the trochanter and moves with it.

4. Tibia – the tibia is the long shank of the leg articulating with the femur by a dicondylic joint so that it moves in a vertical plane. In most insects, the head of the tibia is bent so that the shank can flex right back against the femur.

5. Tarsus – the tarsus, in most insects, is subdivided into from two to five tarsomeres. These are differentiated from true segments by the absence of muscles. The basal tarsomere, or basitarsus, articulates with the distal end of the tibia by a single condyle, but between the tarsomeres there is no articulation; they are connected by flexible membrane so that they are freely movable.

6. Pretarsus – the pretarsus, in the majority of insects, consists of a membranous base supporting a median lobe, the arolium, which may be membranous or partly sclerotized, and a pair of claws which articulate with a median process of the last tarsomere known as the unguifer. Ventrally there is a basal sclerotized plate, the unguitractor, and between this and the claws are small plates called auxiliae. The development of the claws varies in different insect groups. Commonly they are more or less equally well-developed, while in other groups, the claws develop unequally.

All legs are equipped with an extensive arrangement of sensory structures that allow the insect to feel, hear, and taste, providing the insect with its initial assessment of the environment. Chemoreceptors, which are especially prevalent on the basitarsus and eutarsus, provide sensory input on environmental substances and can be used to determine the acceptability of food, ovipositional substrates, and perhaps mates. Mechanoreceptors, most often in the form of hair organs, but also campaniform, chordotonal, and plate organs, provide sensory information on position, movement, and vibrations borne by air and substrate.

Modifications of the basic Leg Structure

The majority of insectan legs are either elongate, slender, and designed for walking and climbing. Various modifications allow the legs to be used in other forms of locomotion and non-locomotory functions. Amongst these are:

1. Cursorial, i.e., adapted for running, as in the cockroach. During walking, the legs form alternating triangles of support, with the fore and hind legs of one side and the middle leg of the opposite side contacting the substrate as the other three legs move forward.

2. Enlarged hind legs (metathoracic legs) of many Orthoptera, fleas, and other insects are saltatorial, meaning they are designed for jumping. The jump of insects such as fleas is aided by a rubber-like protein called resilin in the cuticle that stores and subsequently releases energy for the jump.

3. Fossorial, digging legs, are best known in the Scarabaeoidea and the mole cricket, Gryllotalpa. Powerful, spade-like forelegs (prothoracic legs) of mole crickets, scarab beetles, burrowing mayflies, and other insects are or adapted for digging and rapid burrowing. In Gryllotalpa, the forelimb is very short and broad, the tibia and tarsomeres bearing stout lobes which are used in excavation. In the scarab beetles, the femora are short, the tibiae are again strong and toothed, but the tarsi are often weakly developed. Larval cicadas are also burrowing insects. They have large, toothed fore femora, the principal digging organs, and strong tibiae which may serve to loosen the soil. The tarsus is inserted dorsally on the tibia and can fold back. In the first stage larva it is three-segmented, but it becomes reduced in later instars and may disappear completely.

4. Grasping The ability to hold on is important in ectoparasitic insects. These usually have welldeveloped claws and the legs are frequently stout and short as in Hippoboscidae, Ischnocera and Anoplura. In the latter two groups, the tarsi are only one or two segmented and often there is only a single claw which folds against a projection of the tibia to form a grasping organ.

5. Flattened, fringed legs of aquatic insects such as dytiscid and gyrinid beetles and notonectid backswimmers serve as oars for paddling or swimming (natatorial legs), while long legs with hydrophobic tarsal hairs and ante-apical claws, as seen in water striders (Gerridae), are for skating on the surface of water.

6. Raptorial legs, i.e., those designed to seize prey, have arisen independently in many insectan lineages. Either of the three sets of thoracic legs can be raptorial, but the trait is probably most often expressed in the forelegs (e.g., in Mantodea and Reduviidae) and less frequently in the middle legs (e.g., in some Empididae) and hind legs (e.g., in some Mecoptera).

7. Grooming Insects commonly use the legs or mandibles to groom parts of the body, removing particles of detritus in the process. The eyes and antennae are often groomed, and so are the wings. In Apis and other Hymenoptera there is a basal notch in the basitarsus lined with spinelike hairs forming a comb. Here, the metathoracic legs are modified. A flattened spur extends down from the tip of the tibia in such a way that when the metatarsus is flexed against the tibia the spur closes off the notch to form a complete ring. This ring is used to clean the antenna. First it is closed round the base of the flagellum and then the antenna is drawn through it so that the comb cleans the outer surface and the spines on the spur scrape the inner surface.

8. Pollen collection, The hind legs of Apoidea are modified to collect pollen from the hairs of the body and accumulate it in the pollen basket. Pollen collecting is facilitated by pectinate hairs which are characteristic of the Apoidea. In the honeybee, Apis, pollen collected on the head region is brushed off with the forelegs and moistened with regurgitated nectar before being passed back to the hind legs which also collect pollen from the abdomen using the comb on the basitarsus. The pollen on the combs of one side is then removed by the rake of the opposite hind leg and collects in the pollen press between the tibia and basitarsus. By closure of the press, pollen is forced outwards and upwards on to the outside of the tibia and is held in place by the hairs of the pollen basket. On returning to the nest, the pollen is kicked off into a cell by the middle legs.

9. Silk production Insects in the order Embioptera are unique in having silk glands in the basitarsus of the front legs in all stages of development of both sexes. The basal tarsomere is greatly swollen, and within it are numerous silk glands each with a single layer of cells surrounding a reservoir. There may be as many as 200 glands within the tarsomere, each connected by a duct to its own seta with a pore at the tip through which the silk is extruded.

Other forms of Modifications

Legs also play an important defensive role, not only in permitting escape by running, jumping, burrowing, and swimming, but also in ways such as kicking and slashing. The spines on the legs of many insects, when used in defense, effectively deter predators and competitors and can inflict considerable damage. Insects such as stink bugs and treehoppers deliver powerful kicks at parasitoids and predators that attempt to attack their young.

Autotomy, or the loss of legs at predetermined points of weakness, often at the level of the trochanter, occurs in insects such as crane flies, leaving a predator with only a leg in its clutches as the insect escapes. Legs, whether lost through autotomy or accident, often can be regenerated to various degrees if one or more molts follow the amputation.

In many insects, the legs are used in sound production. Familiar examples include the shorthorned grasshoppers, which have a stridulatory mechanism on the hind femur, involving a series of pegs—the scraper—that is rubbed across a ridged wing vein. Some larval hydropsychid caddisflies have a scraper on the prothoracic femur that is rubbed against a file on the venter of the head. Legs also can be used to produce sound for intraspecific communication by drumming them against a substrate, as in some Orthoptera.

Insects such as Chironomidae seem to use the legs much as a second set of antennae. In Protura, which lack antennae, the forelegs probably have assumed an antennal (i.e., sensory) function.

Chapter 3

Internal Anatomy of Insects

The internal anatomy of insects is made up of various organ systems such as the brain, circulatory system and excretory system. Even though there are a lot of variations between different insects, there are similarities in overall anatomy. The chapter closely examines these key components of the internal anatomy of insects to provide an extensive understanding of the subject.

The Insect Brain

The basic insect nervous system bauplan consists of a series of body segments, each equipped with a pair of connected ganglia, with a paired nerve cord connecting adjacent ganglia in each segment. The ganglia are bulbous structures consisting of neuron cell-bodies and supporting or glial cells and acts as a local processor or computer. the ganglia are interconnected by neurons, constituting a computer network. This plan is variously modified in the various types, but in all cases the ganglia of the head segment form a fused mass, situated above the oesophagus (esophagus) of the gut, and called the supraoesophageal ganglion. This is connected by a pair of nerve trunks (connectives or commissures) that course around the oesophagus on either side and join to the suboesophageal ganglion (SOG or SEG) situated beneath the oesophagus. Many consider only the supraoesophageal ganglion to constitute the insect brain, others (including myself) consider the SOG as part of the brain.

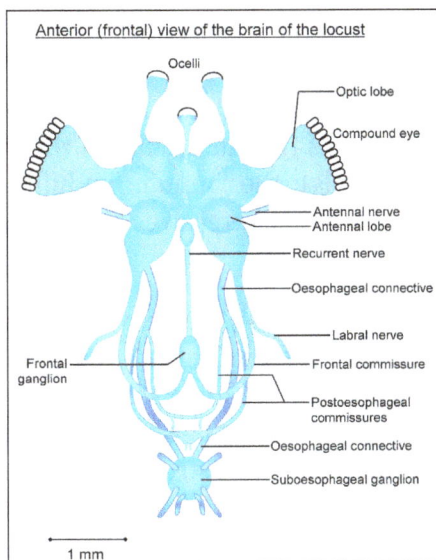

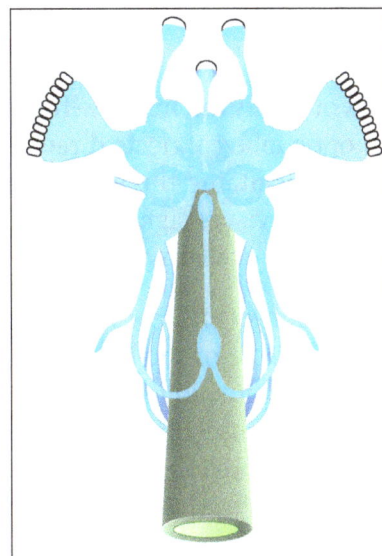

Anterior (frontal) view of the brain of the locust

The relative size of the brain species. That of the diving beetle Dytiscus is about 1/400 of the total body size, that of the ant (Formica) is about 1/280 , and that of the bee about 1/174. The brain is generally larger in those insects that have more complex social lives.

Although much smaller than a human brain, containing only one thousandth as many cells, it is still immensely complex. There is also less replication of function - fewer neurons perform each function.

The supraoesophageal ganglion consists of several fused ganglia or lobes. The paired ganglia of the first (frontmost) head segment form the protocerebrum, concerned with vision, time-keeping, higher functions, memory and combining information from different sensory modalities. Those of the segment segment form the deutocerebrum, which is concerned with processing sensory inputs from the antennae, and also the labial palps and parts of the tegument (body wall). the ganglia of the third segment form the relatively small tritocerebrum.

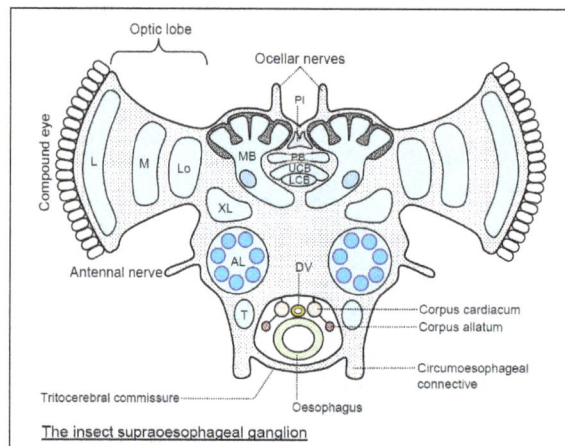

The insect supraoesophageal ganglion

AL, antennal lobe; DV, dorsal blood vessel; L, lamina; LCB, lower central body; Lo, lobula; M, medulla; MB, mushroom body; PB, procerebral bridge; PI, pars intercerebralis; T, tritocerebrum; UCB, upper central body; XL: accessory lobe.

Protocerebrum

The optic lobes of the fly (an insect with particularly good vision) contains about 76% of the brain's neurons. The optic lobe connects directly to the sensory cells (retinula cells) in the retina of the compound eye. It contains three distinct regions (neuropils): the lamina, medulla and lobula, where processing of visual signals begins. The protocerebrum also receives inputs via the ocelli, when present, via the ocellar nerves.

The mushroom bodies (MB, corpora pedunculata, 'stalked bodies') are best developed in social insects, making up 20% of the brain of the bee and 50% of the brain of worker ants (Formica). These are thought to function as higher centres responsible for the most sophisticated computations occurring in the insect brain. Each consists of a topmost cap and a stalk or peduncle (which branches into at least two lobes). The cap consists of a pair of cup-like structures, the medial calyx and the lateral calyx (plural of calyx is calyces). The mushroom bodies receive sensory inputs from the lobula of the optic lobe and from the antennal lobes of the deutocerebrum. Most sensory inputs enter the MB through the calyx. There are about 1000 to 100 000 specialised neurons, called Kenyon cells, in each mushroom body. These neurons have tree-like branching dendrites which receive inputs in the calyces of the MB, a single axon which extends down the stalk of the MB and then gives of branches to two lobes of the MB. Dragonfly mushroom bodies have no calyces and no

Kenyon cells. The mushroom bodies are also involved in learning, and in the honeybee have been shown to process memories, transferring data from short-term memory (STM) into long-term memory (LTM).

The central body receives inputs from the mushroom bodies and integrates sensory inputs from different sensory modalities (such as small and vision) - so-called multimodal sensory perception. It functions as an activating centre, switching on appropriate locomotor activity patterns which are central programs located in the thoracic ganglia. That is it instructs the thoracic ganglia which programs to run - programs that control the legs and wings. These hard-wired programs are sometimes called central pattern generators and require no sensory input for their execution, though sensory inputs may start and stop these programs or modify them slightly.

The pars intercerebralis is a mass of cell bodies, including neurosecretory cells which send their axons to the pair of corpora cardiaca. The corpora cardiac are sometimes fused into a single medial ganglion. They send out nerves to innervate the dorsal blood vessel, forming a cardio-aortic system, which controls the rate of heart beat, as well as having a secretory hormonal function.

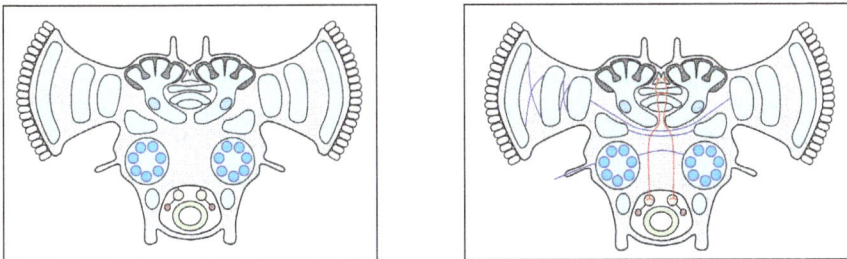

Above: some fibre tracts have been added: tracts in the optic lobe
(notice the crossovers) and connectives between pairs of lobes.

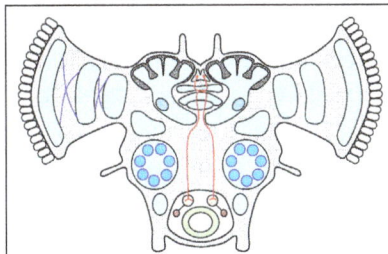

Above: neurosecretory cells with cell bodies in the pars intercerebralis
added. The axons of these neurons terminate in the corpora cardiaca where
they releases hormones: part of the insect neuroendocrine system.

Biological Clocks

Another function associated with the protocerebrum is time-keeping. Insect activity is timed with the daily light/dark cycle - the circadian cycle ('ciracdian' means 'about a day', the exact time being set each day according to environmental cues such as the length of daylight). This timing is due to internal clocks within the insect, which update themselves according to external cues from the environment (zeitgeibers or time-givers) such as the number of hours of light and dark. (This resetting by use of external signals enables the insect to adjust to different local conditions depending, for example, on latitude). Many body parts and organs have their own circadian clocks, indeed each cell appears capable of keeping time, but these appear to be set and synchronised by a central master clock, which resides in the protocerebrum and is both neural and hormonal. In some insects, a master clock is found in each optic lobe, which makes sense as these process light signals. There is also a daily movement of screening pigments in the ommatidia of the compound eye, as the insect adjusts to night-time darkness by increasing the sensitivity of its retina (it will continue to do this at the correct time for days when kept in constant light or dark for several days, so the response is coordinated, in part, by a central clock). Severing of the optic lobes prevents these clocks from synchronising bodily activities. In other species, however, the clock is only abolished if the brain is cut in two, which suggests that it may reside in the central body.

Deutocerebrum

This consists of two nerve centres - the main antennal lobe (AL) and the smaller antennal mechanosensory and motor centre (AMMC) or dorsal lobe. The AL receives inputs from the third (terminal) antennal segment (the flagellum, which is made-up of sub-segments called flagellomeres) via the antennal nerves. It contains from less than 10 to more than 200 sub-centres called glomeruli (singular glomerulus). Inputs to the AL appear to be mainly or exclusively from chemoreceptors (i.e. chemical sensors - olfactory and gustatory, smell and taste) on the flagellum. Each antenna sends signals to the AL on the same side of the head (ipsilateral pathways) although some may also send signals to the AL on the opposite side (contralateral pathways).

Each glomerulus is a region of neuropil (nerve cell processes and synapses) where computations occur. It is thought that each glomerulus may, in some species at least, receive inputs from a specific class of receptor (sensor) on the antenna. For example, in the males of some species there is a specially large glomerulus, called the macroglomerular complex (MGC) which receives inputs from pheromone olfactory sensors on the antenna.

The AL does not receive one input line from each chemoreceptor, as sensors of the same type converge - their axons fuse into a smaller number of axons in the antennal nerve (typically inputs from 15 sensors are combined, a 15:1 ratio). These sensory input axons, and also input axons from the CB of the protocerebrum, synapse with local interneurones within the AL (amacrine cells). Outputs from the AL are carried along the axons of output neurons to the MB of the protocerebrum.

The AMMC receives mechanosensory inputs from mechanosensors (mechanoreceptors)on the first two antennal segments (scape and pedicel) via the antennal nerves. It also sends motor outputs to the muscles of the scape. It also receives inputs from mechanosensors on the labial palps, some tegument (body wall) mechanosensors, and some inputs from the flagellum (possibly from the mechanosensors found on the flagellum). The antennal nerve is therefore a mixed nerve - containing both sensory and motor axons. Some of the antennal mechanoreceptors also send outputs to the SOG, the protocerebrum and the thoracic ganglia.

Tritocerebrum and Stomatogastric System

The frontal ganglion (FG) is an additional free and single (unpaired) median (median = in the midline) ganglion that is connected by a pair of bilateral connectives to the tritocerebrum. A single medial recurrent nerve runs back up to a ganglion situated beneath and behind the supraoesophageal ganglion. This ganglion may be called the stomachic ganglion or the hypocerebral ganglion (HG). In the locust, the HG sends out one pair of outer oesophageal nerves (and one pair of inner oesophageal nerves (ventricular nerves). Each of the latter terminates in a ventricular ganglion (ingluvial ganglion) on the crop of the foregut. These then control crop movements. In Dytiscus, it has been shown that the FG also controls swallowing. Thus, the tritocerebrum and frontal ganglion control the foregut, forming the stomatogastric system. The tritocerebrum also innervates the labrum.

Suboesophageal Ganglion

The suboesophageal ganglion (SOG) and the segmental ganglia of the double ventral nerve-cord each send out pairs of nerves, one of which innervates the pair of spiracles on that segment and so help regulate breathing. (In some insects the segmental ganglia are absent, e.g. in Dytiscus, in which case the lateral abdominal nerves send out nerves to innervate the spiracles). The SOG is a composite ganglion, formed by fusion of the ganglia from the mandibular, maxillary and labial segments of the head and the SOG also sends out nerves to the mouthparts (mandibles, palps, etc.) and so controls feeding behaviours.

The Ventral Nerve Cord

From the suboesophageal ganglion two connectives or nerve cords run back along the ventral side (underside) of the insect. These connect to the thoracic ganglion of the first thoracic segment, T1, which is actually a pair of ganglia, more-or-less fused into a single structure. T1 then gives off two connectives to the second thoracic ganglion, T2 and the sequence continues with a chain of connected ganglia running throughout the length of the insect, in the basic plan. Thus, we say that insects have a double ganglionated ventral nerve cord (VNC). Each ganglion functions as a local processor, regulating the functions of its body segment. The thoracic ganglia are especially well-developed as they have to carry out complex computations to generate patterns of movement in the legs and wings. These output patterns or central programs are contained in the ganglia, but the brain is normally required to switch them on and off. Sensory inputs have little effect on the basic patterns, but do modify them. For example, stress sensors in the wings feedback information to allow fine-adjustments to the wings and control of the angle of attack and wing-twisting. Typically, however, the basic pattern of movement is pre-coded.

The exact arrangement of ganglia is grasshopper, left, the first three abdominal ganglia (A1, A2 and A3) are fused with the third thoracic ganglion (T3) to form a composite ganglion. The final abdominal ganglion is often composite. It is debatable how many abdominal segments there are, as the last few are modified and reduced, but generally there are 9-10, and the ganglia of these segments are fused with that of A8, again forming a composite ganglion ('A8'). Each ganglion gives off nerves to the various structures in its body segment. However, complications arise as ganglia may receive inputs from certain other segments too.

In flies ganglia are highly fused. Typically the three thoracic ganglia and the first four abdominal ganglia are fused together, into a single ganglion in the thorax. The remaining 5 (or so) abdominal ganglia are also fused into a single abdominal ganglion. The connectives of the nerve cord between the composite thoracic and composite abdominal ganglia then give off pairs of nerves to those abdominal segments lacking a regional ganglion. Such fusion of ganglia concentrates processing power where it is needed and reduces the time wasted by sending signals up and down the nerve cord between ganglia that may need to cooperate.

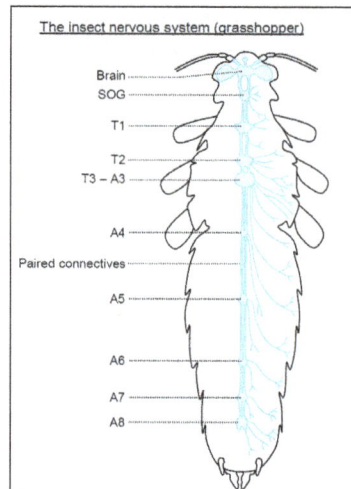

The insect nervous system (grasshopper)

Learning, Memory and Intelligence in Insects

Although the mushroom bodies of the brain have been shown to be involved in learning, ganglia other than the brain are also capable of learning. Learning has been demonstrated in decapitated cockroaches! If a headless cockroach is wired so that one of its legs receives an electric shock when lowered, then it will learn to avoid the shocks by keeping the leg raised! (This is a classic experiment). The thoracic ganglia are responsible for controlling leg movement, and it is these ganglia that learn the new behaviour. Intact cockroaches have a preference for darkness if given a choice between the illuminated half of a chamber and the darkened half. However, if they receive electric shocks in the darkened half, then they will learn to avoid the dark-half and remain in the light.

Habituation. An isolated cockroach leg exhibits another phenomenon related to learning - sensory habituation. If a touch-sensitive bristle is stimulated on an isolated leg and the activity in the leg nerve recorded, then it will be found that the strength of the stimulus diminishes if the stimulus is repeated rapidly, or sustained. This is due to fatigue in the periphery nerve and sensor, and this ensures that insects respond to changes in the environment, and learn to ignore persistent stimuli that are of no relevance. For example, body lice prefer rough fabrics, like wool, to smooth fabrics like silk. When crossing from wool to silk they will keep turning as they try to find their way back onto the wool. However, if they fail to find the wool (say if it is removed) then this behaviour stops as the lice learn to make do - they habituate. Habituated insects are still responsive to changes in the stimulus, however. Thus, if the texture of the fabric for our louse on silk changes again, then we expect it will respond again, and habituate again if necessary.

Learning one's way about - route learning. Some insects, such as ants and cockroaches, are capable of learning the route of a maze, if the maze exit leads to reward, or if escaping from the maze avoids punishment. A cockroach will navigate a maze to it home-pot, so long as the home-pot contains recognisable cockroach odours. On subsequent trials it will reach its home-pot with increasing ease as it gradually learns the route. When compared to rats, ants learn mazes at half the speed: a rat will master a maze after about 15 runs, an ant after about 30, though the ants still make a few errors. (Not bad when you consider how tiny the ant brain is!). However, if a change is made to the maze, such as reversing the pattern, then rats learn the 'new' maze more rapidly than before, recognising the similarity in the patterns and transferring their previous learning to the new situation (transfer learning). In contrast, the ant starts all-over again, treating the maze as entirely new. Thus ants have little or no capacity for transfer learning, that is they seem unable to apply what they have learnt to a novel situation.

One of the most impressive feats of insect learning is locality learning. In addition to learning routes, insects can recognise the locality in which their nest is situated, or in which food is found. This involves exploratory learning- the insect will typically fly around a bit after leaving the nest, learning the position of many landmarks very rapidly and then leave. This is latent learning, meaning that a period of time elapses between learning and reward. The reward occurs when the insect returns home and locates its nest. This can be demonstrated by experiment. For example, in classic experiments on the beewolf, Philanthus triangulum a type of digger wasp, which brings back food to its developing young in the nest, the wasp will learn to recognise a ring of pine cones placed around the nest entrance and that the nest hole is the centre of this ring. If the pine cones are displaced a few centimetres when the wasp leaves, however, it will return to the centre of the ring of pine cones (only its nest is no longer there!).Generally, however, these insects are only temporarily confused by changes to one or two landmarks, falling back on other landmarks further from the nest (and perhaps other cues like smell?).

Associative learning occurs when a stimulus, irrelevant by itself, is made relevant by pairing it with something meaningful, like a food reward and the animal learns to identify the hitherto unconnected stimulus with food. For example, a cockroach can be rewarded by being given food when presented with a particular, but unrelated, odour. It will then learn to associate that odour with food and be attracted to it when foraging. Bees can be trained to associate a particular colour or pattern 9though their perception of shape is limited) with food.

Short-term and long-term memories. When an insect has just completed a task, learning is abolished if a new task is undertaken immediately after. Learning requires a latent period of rest in-between activities. To some extent the same is true of humans - learning becomes greatly enhanced if breaks are taken every 20 minutes or so during study. It is during these rest periods that the brain processes the information (often subconsciously in humans) and the appropriate neural pathways are reinforced. [Dreams in humans are especially curious - if activities are undertaken straight after waking then dreams are often very easily forgotten, but if a few minutes are spent reflecting on them they may be remembered more easily]. This can be explained by the existence of short-term memory (STM) and long-term memory (LTM). The nervous system requires time to process information in short-term memory, making sense of it and discarding information deemed irrelevant and transferring more useful information into long-term memory. In insects LTM typically retains information for several days.

Do Insects Sleep? Hard to say. Insects certainly exhibit circadian cycles. Insects enter a dormant or semi-dormant state when chilled. This may occur seasonally, during winter, or daily at nightfall. At nighttime insects may assume a particular posture in which to rest, bees may hold onto vegetation with their tightly clamped jaws, for example. They may also return to the same resting spot each night. During these quiescent states the body temperature drops and energy is conserved. Interestingly, learning increases when insects are given rest between tasks or training sessions, as already mentioned, but also if they are chilled during these rest periods. Could it be that when chilled into torpor at nighttime insects are forming memories? It has been suggested that two chief functions of sleep are: energy conservation and memory formation. In this case the insect nighttime torpor is not so very different. Many insects are, of course, nocturnal. Cockroaches are nocturnal and learn better at night than during the day. It has been suggested that this encourages them to remember useful things like lessons learned during nighttime foraging.

Temporal learning occurs in insects. As we have seen insects can measure time, by the use of internal body-clocks, and they can also compensate for the changing position of the Sun in the sky when they use the Sun to navigate. To navigate successfully by the Sun they must know what time it is, in the sense of what portion of daylight has passed. Observations have shown that this may require learning, with young insects making mistakes by assuming that the Sun stays fixed in the sky.

Insight learning is a higher form of learning, similar to transfer learning, in that it takes prior learning and applies that to a new problem. One example is tool use. Some insects use tools, for example, some digger wasps fill in their nest burrow once the young is mature and ready to pupate (and so requires protection but no more food). The female may then hold a small 'pebble' in her mandibles and use it to pound-down the earth, and then discard the pebble. However, this behaviour is not intelligent in the sense that the insect reasoned a solution, rather it is inherited genetically and is instinctive and so not true insight learning.

The Peripheral Nervous System

Peripheral nerves may be sensory or motory, but in insects are generally mixed. For example, the antennal nerve carries many sensory fibres conveying inputs from the many antennal sensors to the brain, but it also contains some motor fibres carrying output signals to the muscles in the base of the antenna. Animal nerve fibres, which are usually the axons of neurons, are typically wrapped by insulating glial cells. In nonmyelinated axons, the wrapping may be a simple sheath that loosely invests a group of axons. One function of this sheath is to ensure that the axons are bathed in a suitable salt solution necessary for them to conduct impulses. In myelinated axons, however, each individual axon is tightly wrapped in its own insulating sheath of material called myelin, which folds tightly around the axon several times, except at exposed regions (such as the nodes of Ranvier in mammals, although these may be simple pores in other animals). This more advanced type of wrapping is insulating and serves to speed-up nerve transmission. Students of vertebrate zoology often mistakenly believe that only vertebrates possess myelinated axons (in addition to unmyelinated axons which also occur in vertebrates). Myelinated axons occur in many invertebrates. Indeed, the fastest nerve transmission in the animal kingdom is seen in the myelinated axons of certain shrimps, which conduct signals at about 200 m/s. Insects have an intermediate form of insulation, which is like the myelin sheath, except that the myelin is wound loosely, leaving fluid-filled spaces between the layers. Such a nerve fibre is intermediate between unmyelinated and myelinated and is called a tunicated nerve fibre.

Insects Locomotion

Insects have been described as 'ideal miniature robots'. Indeed, the way they move is extremely efficient and is being used as a basis by robotocists to develop locomotion in their own machines.

The locomotive power for insects comes from the thorax and its appendage (legs and wings).

Insect locomotion is characterised by two main advancements: the jointed limbs characteristic of arthropods that allowed the arthropods to colonise land and flight. The jointed limb of arthropods (arthropod meaning 'jointed leg') evolved in marine forms, such as crustaceans that are often amphibious, for example many shore crabs. The tough exoskeleton supports the weight of the animal and also has internal projections to which the muscles attach. If you have ever eaten a crab claw, then you may have noticed the tough 'tendons' that attach to it inside and which can be pulled like levers to operate the claw. The muscles attach to these tendons which are internal extensions of the exoskeleton (and really constitute an internal skeleton). The joints permit the rigid exoskeleton to move and consist of softer articulating membranes.

The arthropods were the first animals on Earth to truly colonise dry land. The jointed exoskeleton could support their weight and the cuticle, with the addition of waterproofing waxy layers prevented dehydration. The first land arthropods were myriapod-like animals: centipedes and millipedes, some as large as a human, living in the forests of mosses. Like myriapods today they had many body segments, each bearing a pair of legs, whilst the head developmentally was equivalent to several segments fused together, with the limbs adapted as palps for feeling and tasting, antennae for smelling and mouthparts for eating. Over time, some of their ancestors lost many of their legs as locomotion became refined, making do with just 8, as in arachnids, or 6 as in insects (hexapods). Archaic insects are intermediate, Marchantia, for example, having never evolved wings and still bearing rudimentary legs on the other segments and with palps that still look like a 4th pair of legs.

Insect Legs

The arthropod leg can be divided into a number of segments, or podites, starting at the base and working towards the foot these podites are: the cox, trochanter, femur, tibia, the tarsus composed of five segments or tarsomeres, and finally the pretarsus, present as a pair of claws and sometimes also an adhesive pad. These segments are variously modified in different insects. In the honeybee hindleg (metathoracic leg) for example, the tibia contains the pollen basket for carrying pollen and the first basal-most tarsus segment, the basitarsus, the pollen brush.

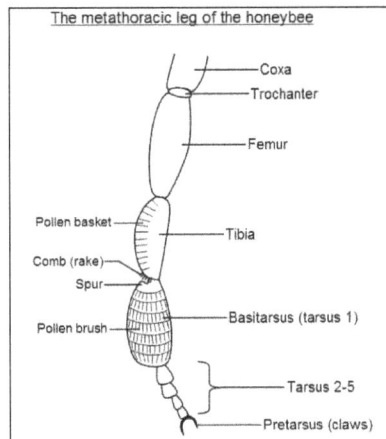

The metathoracic leg of the honeybee

Coxa
Trochanter
Femur
Pollen basket
Tibia
Comb (rake)
Spur
Basitarsus (tarsus 1)
Pollen brush
Tarsus 2-5
Pretarsus (claws)

The honeybee uses its hindlegs to collect pollen. When it forages for nectar and pollen, pollen sticks to its hairy body. In flight it grooms itself with its hindlegs, which have special rows of bristles forming the pollen brush on the basal tarsal segment, which is enlarged. It then cleans the brush with a comb of bristles situated at the bottom of the tibia, the left tibia combs the right leg pollen brush and vice versa. A projection on the top of the basal tarsal segment, forming a spur, then pushes the pollen from the comb into the pollen basket, which is also on the tibia.

In aquatic insects the legs may be modified into paddles or ores for swimming, and in the crustaceans we see a bewildering array of modified limbs for performing different functions. The exoskeleton makes the arthropod limb infinitely adaptable.

Typically the coxa and trochanter are short segments, and operated by muscles that insert inside the thorax (extrinsic leg muscles). The tibia is operated by muscles in the femur above it (intrinsic leg muscles). The femur itself generally moves little, being more-or-less fused to the trochanter. In a jumping insect, such as the grasshopper, the femur is visibly greatly thickened and enlarged to accommodate the powerful jumping muscles that operate the tibia.

Jumping Insects

A grasshopper can high-jump vertically upwards 45 cm (10 times its body length) and long-jump 90 cm horizontally (20 times its body length). This is equivalent to a human high-jumping 60 feet (20 m) and long-jumping 120 feet (40 m)! Insects are also strong for their size, being capable of carrying 2-3 times their body weight, and up to 7-10 times their body weight in the case of ants. (Mind you human powerlifters can comfortably lift 2-3 times their body weight or more, but they probably could not carry it far). This great strength is actually due to their small size and it is unlikely that a grasshopper the size of a human could really jump 60 feet! The strength of a muscle increases with the cross-sectional area or girth (thickness) of the muscle, that is with the square of linear dimension, whilst weight increases with the cube of length. Thus if we double the height of a human, keeping all proportions and other factors constant, then although their strength would increase 4-fold, their weight would increase 8-fold and so their strength : weight ratio would actually half! Thus, the great strength of insects is easily explained. In some cases an insect muscle may consist of only one or a few muscle cells, as they don't need much motive force to effect great strength for their size.

The thrust, or more specifically the force to weight ratio, of a grasshopper's jumping leg is about four times that generated by a human leg. (In comparison, measurements indicate that the jumping power of a chimpanzee leg is equal to that of a human leg, even though the muscle volume of the chimpanzee leg is about half and chimpanzees typically lighter - chimpanzees are very good jumpers). This can be partly explained by the greater relative strength of insects, due to their small size, but this results in an acceleration rate that muscles cannot achieve unaided. It is the fact that the insect cuticle contains a very elastic material called resilin, which is abundant in the cuticle of the legs of jumping insects, that enables the insect leg to store energy. When preparing to jump the leg is locked in place as the powerful tibial flexors (situated in the femur) contract, straining the cuticle of the locked leg which builds up elastic energy. Suddenly, the cuticular lock is released and the leg springs straight, releasing the stored elastic energy as the insect jumps. Resilin is almost perfectly elastic: 97% of the stored energy is released as mechanical energy, only 3% is lost as unusable energy (heat and sound).

Insect Flight

The apterygotes are archaic insects that never evolved wings and are a kind of living fossil. A familiar example is the domestic silverfish, which you may find scurrying around in kitchens and pantries, looking for crumbs or bits of wood and paper to eat. Their fish-like serpentine motion and silvery bodies can make them look like tiny fish swimming across the floor! This snakelike movement of the body increases their stride-length and speed (since speed = stride length x stride rate). Amphibians deploy a similar tactic.

The anatomy of the flight muscles is illustrated below:

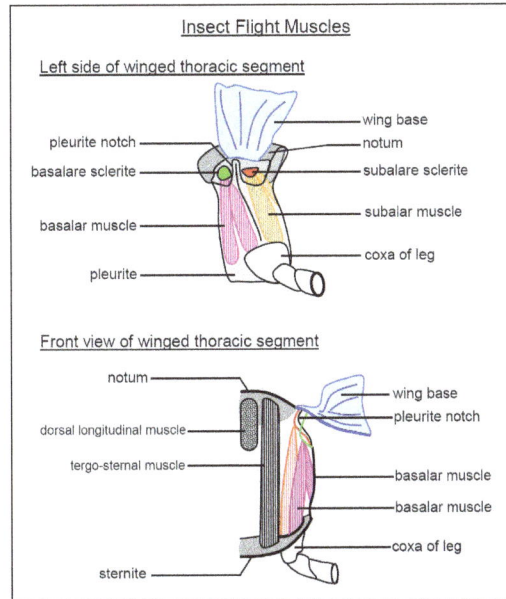

Each thoracic segment is encased in four groups of cuticular plates: the notum (tergum) on the back, the sternum on the front and the pleura (singular pleuron) on the sides. These four regions of the cuticle are made up of cuticular plates, called respectably the tergites, sternites and pleurites, and softer artioculating membranes that connect the regions together. The flight muscles attach to certain of these plates or sclerites.

There are two sets of flight muscles - direct and indirect muscles. The direct muscles attach to the wing base, or more specifically to sclerites (cuticular plates of the exoskeleton) in the pleura (the groups of cuticular plates making up the sides of the insect) that then attach to the wing base. The base of the wing attaches to the notum and pleuron. The notum (or tergum) is the plate that forms the insect's back and the pleuron its side.

Direct muscles. The direct muscles attach to sclerites in the pleura at both ends. There are two paired sets of these muscles: the basalar muscles and the subalar muscles. The wing hinge pivots on a projection of the main pleurite, called the pleurite notch or pleural wing process (PWP). The basalar muscles attach just anterior of this pivot point, to a small pleurite called the basalare sclerite. The subalare muscles attach just posterior to this pivot, to a small pleural sclerite called the subalare sclerite. In dragonflies, the wings do not need to move very fast, as these insects are large, and these direct flight muscles provide the motive power, lowering the wing by contracting. However, in most insects the direct flight muscles are only used to tilt the wing. The wing root contains a number of small cuticular plates, called axillary sclerites, the second axillary sclerite articulates with the PWP and connects to the subalare. the subalare is hinged to the main pleurite just behind the PWP, the basalare in front.

When most insects fly, the wing tips trace a figure-of-eight - tipping the leading edge downwards as the wing moves forwards and downwards. This is achieved by contracting the basalar muscles, pulling on the basalare sclerite, depressing it and transferring the strain to the wing base, due to the elasticity of this region of the cuticle (due to resilin again). This forward-downstroke is the main power stroke. The wing begins the downward loop with a high angle of attack, but as the leading edge tilts downwards, the wing momentarily becomes horizontal, in the middle of the stroke, minimising the angle of attack. At this point the wing is moving fastest, and reducing the angle of attack prevents stalling. When the wings move upwards and backwards, during the recovery stroke, the leading edge tips backwards; this is achieved by contracting the subalare muscles, lifting the leading edge of the wing about its pivot. The wing is rotated again at the top of the recovery stroke, restoring the maximum angle of attack just before the downstroke begins.

Indirect muscles. These provide most of the motive power in most insects, whilst the direct flight muscles rotate the wings, controlling the angle-of-attack. They consist of two antagonistic pairs of muscles: the dorsal longitudinal muscles and the sterno-tergal (or dorso-ventral) muscles. [These muscles occur in most segments of the insect body, thoracic and abdominal, where they allow the segments to change shape and telescope in and out of one-anotehr to some degree, but are especially well-developed for flight in the meso- and metathoracic segments, that is the second and third segments which bear the wings).

The dorsal longitudinal muscle pair run lengthwise along the top (dorsal side) of each (winged) thoracic segment, attaching to the phragmata, which lie just beneath the tergal antecostal sutures, at both ends. The antecostal suture is an intersegmental ring (visible from the outside as the junction between segments), where the tergite (and sternite) curve inwards at the ends of each segment. The tergite of each segment curves down, underneath the back of the tergite in-front of it (rather like subducting tectonic plates) at the front. This forms an internal cuticular ridge, running transverse (across the thorax) at the front of each segment, called a phragma (plural phragmata). In each segment the dorsal longitudinals attach to the phragma of its own segment at the front, and to the phragma of the segment behind it at the back. Contraction of these muscles will shorten (and deepen) the thoracic segment slightly. This raises the notum, causing the wings to pivot on elastic hinges and move downwards (wing depression).

The tergo-sternal muscle pair (or sterno-tergal or dorso-ventral muscles, etc.) run vertically, joining to the sternite at the bottom and to the tergite at the top. Contraction of these muscles flattens the thoracic segment (and lengthens it) slightly. This pulls the notum down slightly, causing the wings to pivot on elastic hinges and lift upwards (wing elevation).

In many insects, the notum actually switches between a depressed state and a raised state quite suddenly, as the cuticle clicks in place - the so-called click mechanism. This means that as the indirect flight muscles contract, strain builds up in the elastic cuticle, storing energy that is released when the notum suddenly click into a new position, releasing the built-up elastic energy which helps power the wings. Summary of muscle action:

- During the downstroke: the tergo-sternal muscles relax as the dorsal longitudinals contract, depressing the wing and generating lift as the wings move forwards and downwards. Basalar muscles contract, as the subalar relax, tilting the leading edge of the wing downwards (to prevent stalling).

- During the upstroke: the dorsal longitudinal muscles relax as the tergo-sternals contract, lowering the notum and raising the wings during the recovering stroke as the wings move backwards and upwards. subalar muscle contract as basalar muscles relax, raising the leading edge of the wing.

Insects can achieve flight speeds of up to 25 mph (40 kph). A dragonfly, for example, can cruise at 18 mph, a honeybee at 14 mph. A dragonfly will beat its wings at 20-30 beats per second (bps or Hertz, Hz), a housefly (Musca) at 100 bps, a mosquito at 300-600 bps and Forcipomyia at 1000 bps, which is extremely fast! The faster of these exceed the flicker fusion frequency of the human eye (50-60 Hz) at which motions blurs, and so the individual wing beats become impossible to see without playing slowed down recordings.

Control of the Flight Muscles

Nerves can conduct impulses no faster than about 30 pulses per second. How then can these slow nerves control such fast muscles? In the dragonfly, there is no problem, one nerve impulse can bring about one wing beat - the flight muscles are controlled neurogenically. The muscle contractions and nerve pulses are synchronous. In insects that beat their wings much faster, however, nerve impulses simply start and stop the machine, once the wings get going the indirect flight muscles are alternately stretched as they relax - the dorsal longitudinals stretch when the tergo-sternal contract, and vice versa. When these muscles are stretched, this stretching triggers their subsequent contraction, so cycles of relaxing-stretching-contracting can occur rapidly, without having to wait for the nervous system to keep up! The nerve impulses and muscle contractions are then asynchronous.

Flight Muscle Physiology

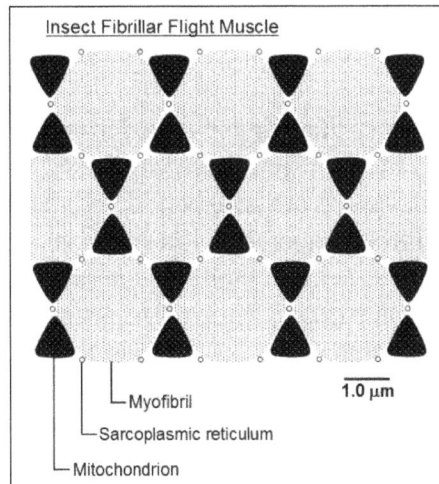

Insect Fibrillar Flight Muscle

Myofibril
Sarcoplasmic reticulum
Mitochondrion
1.0 µm

The direct flight muscles are very well-developed and resemble vertebrate muscle in many ways. The indirect flight muscles, however, are exceptional! They are extremely powerful. Vertebrate biologists may have heard that hummingbird flight muscles are the most powerful in the animal kingdom, power referring to work rate, this is partially correct. In fact they tie for the title of most powerful muscles with insect indirect flight muscles! Both muscle types have very similar power outputs. (Old reports that insect flight muscles could generate the power output equivalent of a piston-driven combustion-engine were erroneous, however.) These indirect flight muscles have very characteristic structure under the electron microscope. Not only are they striated, like vertebrate muscles, but each cell or muscle fibre is packed with contractile motor proteins and has a well-developed sarcoplasmic reticulum (membranous sacs that store calcium ions which signal and initiate contraction) and a regular array of well-developed mitochondria. The whole looks remarkably regular and semi-crystalline, as illustrated below. They are thus called fibrillar muscles.

The Role of Vortices in Insect Flight

Insects do not generate all their lift and thrust simply by pushing air downwards and backwards as they beat their wings. The insect wing also acts as an aerofoil as its lices through the air (which is why angle of attack is so important). This means that as air flows over its surface it is set into circulation around the wing. The inevitable result is that vortices are shed from the tips of insect wings and these vortices (or rather the circulation that causes them) are essential in lift generation.

Other Flight Mechanisms

Some of the smallest insects have wings that are not aerofoils, but simple groups of bristles! These seem to work more like oars, rowing the insect through the air. This happens because for a very small insect the Reynold's number is very low. Reynold's number is the ratio of inertial forces (due to bulk fluid flow) to viscous forces (due to the 'stickiness' of a fluid). At very small scales viscous forces dominate and air becomes more like water (and water like treacle).

The tiny wasp Encarsia formosa generates lift by a clap-and-fling mechanism. The insect alternately claps its wings together behind its back, and then flings them apart. When they part, they separate first along the leading edge, causing air to rush into the space between them. This causes air to circulate around the wings, again generating vortices and lift, but without the normal aerofoil mechanism.

The Insect Wing

Evidence from the fossil record indicates that wings evolved from lateral outgrowths of the nota (tergites) of the thoracic segments. It is thought that these were used to absorb heat, perhaps when the insect as basking in the sun at dawn. Insects are largely poikilothermic ectotherms, meaning that they do not regulate their body temperature by generating internal heat (though this is a generalisation and some insects vibrate their wing muscles prior to flight to generate the necessary heat), in other words they are 'cold blooded'. Absorbing heat through these plates which may have been well-supplied with blood and have a large surface area would warm the leg muscles of the thorax, ready for action. They may also have been used in respiration, absorbing oxygen for the leg muscles. They could also have been used in visual displays, perhaps in courtship and may have been brightly coloured.

It would appear that the wings of modern insects can still perform these functions. The wings are thin membranous 'gossamer' sheets of little more than a double layer of epithelial cells covered by chitinous secretion. This structure is supported by a series of veins, each containing a trachea and perhaps a nerve bathed in haemolymph (insect 'blood') which runs in the channel between the trachea and the epithelium wall. There is evidence that insect blood facilitates the transport of oxygen, though this is mainly carried out by the tracheal system and the haemolymph often runs in channels close to the tracheae. The large wings of lepidopterans are thought to have a significant role in absorbing oxygen from the air. Beating the wings must assist this by circulating the air around the wings.

The pattern of veins (venation) of the insect wing varies considerably among species, but there is a general pattern. Along the leading margin (costal margin) of a wing runs a prominent vein called the costa. Behind this is the subcosta and then a pair of veins that branch from a common source, called the radial and medial veins. Behind this is a separate group of veins called the anal veins (as they span the rear portion of the wing).

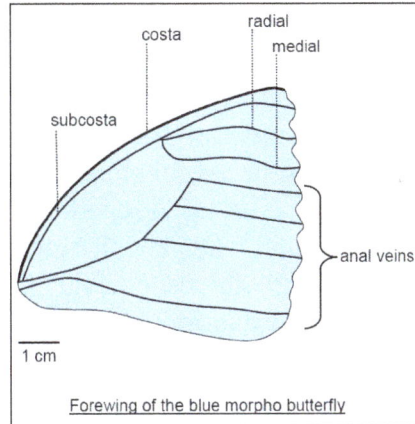

Forewing of the blue morpho butterfly

Insect 'blood', properly called haemolymph, circulates around the wing veins, maintaining the wing which would otherwise become dry and brittle. Accessory hearts or muscular pumps are often present either at the root of the veins at the wing base or along the course of the veins, helping to pump the haemolymph around the wing. The main pumps, however, are located in the thorax. The haemolymph typically enters the wing at the costa, travels around the margin of the wing and then returns along the more posterior veins, such as the anal veins. Remember that insects have an open circulatory system, so there are regions in which haemolymph flows rather loosely in haemolymph spaces rather than vessels, as between the costa and anal veins.

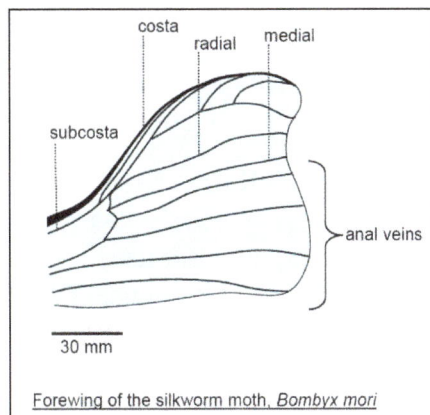

Forewing of the silkworm moth, *Bombyx mori*

The wing is equipped with sensors. Bristles, presumably tactile, typically line the margin, and clusters of tiny dome-shaped campaniform sensilla. A campaniform sensillum consists of a small cuticular dome, sometimes at the bottom of a shallow pit, with a mechanosensitive dendrite underneath. The dendrite is a process from a modified epithelial cell. The number and arrangement of these sensilla depends on species. They are usually found on the undersurface of the wing. In the locust, Schistocerca, there are about 95 on on the undersurface of the costa, one group of about 20 near the wing base (arranged in a fan-shaped array) and the second group of about 60 towards the wing tip.

The cockroach has only 20-40 campaniform sensilla, mostly on the forewings, whilst good fliers, like true flies, bees and wasps, may have 6-7 times as many as the locust, arranged in 6 groups or so. The campaniform sensilla are proprioceptors, detecting flexing and torque in the wing cuticle. Strains in the wing cause the domes to reversibly buckle, activating the sensory dendrite underneath. They can thus detect changes in position of the wings. Their domes are elliptical, rather than round, and they are aligned in such a way that they are most sensitive to wing twisting (pronation and supination) such as when the wing changes its angle of attack through the wing stroke-cycle.

The main pattern of wing movement is generated by the central nervous system. Each segment of the insect body has a pair of ganglia that are more-or-less fused into a single ganglion, situated ventrally and connected to the ganglia in other segments by the double ventral nerve cord. The ganglia of some adjacent segments may also be fused together, especially in the head and abdomen. These act as local processors or 'brains'. The ganglia of the mesothoracic and metathoracic segments, the wing-bearing segments, are chiefly responsible for controlling the wings. The brain is not directly required. Some of the anterior abdominal ganglia may also contribute. These form a central-pattern generator that generates the correct pattern of firing to move the wings even in the absence of sensory input. However, sensory input, chiefly from the campaniform sensilla of the wings is essential to maintain the correct frequency of wingbeat and also in the control fine movements of the wing, such as wing twisting.

The antennae are also involved in regulating flight speed. Both the strain acting on them by the moving air, and the vibrations this causes, is sensed by Johnston's organ in the base of each antenna, giving a measure of air speed. Initiation of flight can be brought about by loss of contact between the feet (tarsi) and the ground, by pinching the abdomen, but this soon stops unless air is kept blowing on sensory bristles on the insect's face. Thus, the head may not be needed to initiate flight, but it has a role in maintaining it and in regulating speed.

The importamce of feedback from the sensory organs is illustrated by the observation that many insects can continue to fly with one of their four wings missing. Dragonflies are especially noted for this, being able to compensate and maintain stability in the air despite losing one of their hindwings, though their speed will be reduced due to a loss of thrust.

The hindwings typically provide most of the thrust, each hindwing providing about 35% of the thrust in a flying locust. The forewings may be toughened top provide protection for the hindwings, forming parchment-like tegmina (singular tegmen), or in the case of beetle the very hard elytra (singular elytron) or wing-covers. In many true bugs, Hemiptera, only the basal part of the forewing is thickened and these are called hemelytra. In true flies, Diptera, the hindwings are reduced to the small club-like halteres, which beat out of phase with the forewings and act as sensory gyroscopes to maintain flight-stability (they are loaded with campaniform sensilla).

It is important that the forewings do not interfere with the hindwings during flight. In locusts, cocockroaches and dragonflies the four wings are kept separate but each pair beats out of phase with the other. For example in the locust the wings beat about 17 times per second and the forewings lead the hindwings by about 6-8 millseconds. Some insects solve the problem by coupling the wings together. For example some moths and butterflies have jugal coupling, in which a toothlike process extending from the hind-margin (anal margin) of the forewing, called the jugum, fold under the hindwing in flight, so that forewings and hindwings are locked together and function as a single pair of wings. This is much like cutting a slit in the margin of a sheet of paper and slotting another sheet into it. In some moths, the mechanism used is frenulo-retinacular coupling, in which a spinelike frenulum on the fore-margin (costal margin) of the hindwing hooks under the loop-like retinaculum on the underside of the forewing. Amplexiform coupling, seen in some butterflies, involves no special hooks, but the wings overlap significantly.

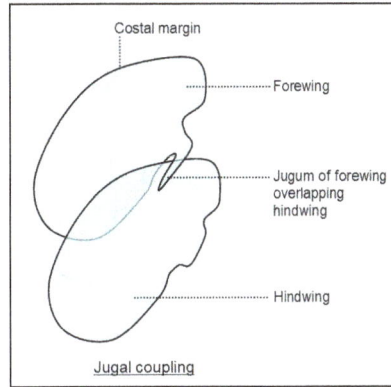

Walking and Running in Insects

The standard explanation of insect walking is as follows. The insect maximises stability by keeping a supporting tripod of three feet on the ground at any one time, the front and hind-leg of one side and the middle leg on the other, whilst the other three are lifted in the air stepping forward to form the next tripod. With each step the insect falls from one tripod to the next, such that there is an alternation from a tripod formed by the fore and hind-legs of the right side with the middle leg of the left side and a tripod formed by the middle leg of the right-side and the fore and hind-legs of the left side and so on. In each step the hindleg on the ground provides most of the thrust, causing the insect to pivot to the side. (When the right hindleg is pushing on the ground it pivots to the right, and when the left hindleg is pushing it pivots to the left). Thus, the insect moves in a zig-zag path as it advances. Keeping at least three feet on the ground at any one time gives optimum stability, as a tripod is a very stable structure, preventing the insect from falling over. This stability is important as insects, despite offering less surface to the wind, can be blown a few millimetres off course by strong winds and so they have to make sure they are do not fall. This stability minimises on central nervous system processing - in bipeds, for example, the brain has to do much more processing to ensure that balance is maintained.

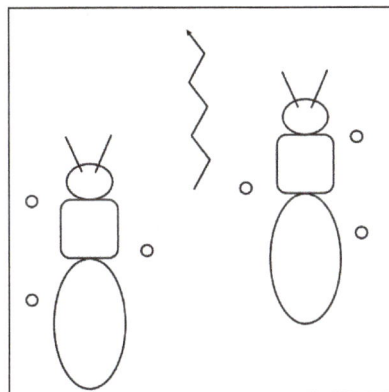

Above an insect taking a 'step' with the circles
showing those feet in contact with the ground.

However, you may not be surprised that this is a simplification. Some insects show a different walking pattern - they keep the tripod but lift the other three feet in sequence, one at a time, so that 4 or 5 feet are in contact with the ground at any one time. This increases stability and also prevents pivoting so that the insect can move in a straight line, as illustrated below:

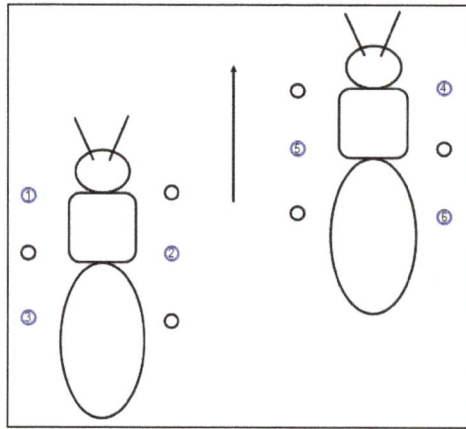

An alternative walking pattern in some insects. The three numbered blue circles represent those feet being lifted from the ground, the numbers indicating the order in which they are raised.

Generally speaking, having fewer feet on the ground at any one time increases stability but reduces speed. Thus some insects will switch their walking pattern, keeping fewer feet on the ground at higher speeds. When they run, some insects have only two feet on the ground at any one time.

Crawling in Insects

Insect larvae may be capable of walking, but many crawl. In addition to the six thoracic legs, many insect larvae can protrude pseudo-legs or prolegs from their abdomen. In the caterpillars of some moths and butterflies a wave of muscular contraction travels from the rear to the front, with the prolegs and then the thoracic legs advancing forward in turn, beginning with the rearmost, as the wave reaches them. The haemolymph provides hydrostatic pressure, so that the larva is pressurised, and when muscles in the body wall of a segment contract this pressure increases, protruding the prolegs by hydrostatic pressure. (The prolegs are essentially inflatable). Specialised body wall muscles maintain the internal pressure. Inchworm or looping movements may assist: arching of the caterpillar helps bring forward the prolegs as the rear advances, and then straightening out of the body advances the front section bearing the thoracic legs.

One of the characteristic features of insect muscles, especially those in larvae, is supercontraction. In vertebrate muscles, each muscle fibre contains myofibrils, bundles of protein fibres: actin and myosin filaments parallel to its long axis. Mobile, lever-like cross-bridges form between the myosin and neighbouring actin filaments when the muscle contracts, pulling the filaments over one-another and shortening the muscle. Along the length of each myofibril, the muscle is divided into sarcomeres, each sarcomere containing an array of actin and myosin and separated from neighbouring sarcomeres by z-bands (z-lines or z-discs), one z-band at each end of the sarcomere. The actin filaments are joined to one z-band at one end, but the myosin filaments are not attached and occupy the centre of each sarcomer. The actin (thin) and mysoin (thick) filaments slide past one-another as the muscle shortens. The z-bands thus place a limit on how much the muscle can contract - once the myosin filaments hit the z-discs as the sarcomere shortens, which happens when the sarcomere length equals the myosin length,the muscle can contract no more. Vertebrate muscles can contract no more than 50%. In many insect muscles, however, the z-bands are perforated, with pores through which the ends of the myosin filaments can pass. Thus the sarcomere can shorten much more, to a length shorter than the length of the myosin filaments. These muscles can shorten by 76%, a phenomenon called supercontraction. This supercontraction partly explains the great flexibility of the bodies of many insect larvae, assisting their locomotion.

Another curious feature of insect muscles is that they have more actin filaments surrounding each myosin filament. In vertebrate muscle, each myosin is surrounded by six actins and each actin by three myosins (a 2:1 actin : myosin ratio). In most insect flight muscles each mysoin is also surrounded by six actins, but each actin is adjacent to only two myosins (a 3:1 ratio). In some insect leg muscles there may be as many as 12 actins around each myosin. Initially this was thought to increase contractile force (power and strength) by enabling more cross-bridges to form. However, measurements show that insect muscle is similar in strength to vertebrate muscle, per unit of cross-section, sometimes slightly less. To be certain one would have to measure speed and power generation in addition to strength - these three parameters are not equivalent. At the moment the reason for these differences between insect and vertebrate muscle are not certain, but the greater actin : myosin ratios in insect muscles are correlated to greater overlap between the actin and myosin filaments and it is thought that they are related to supercontraction.

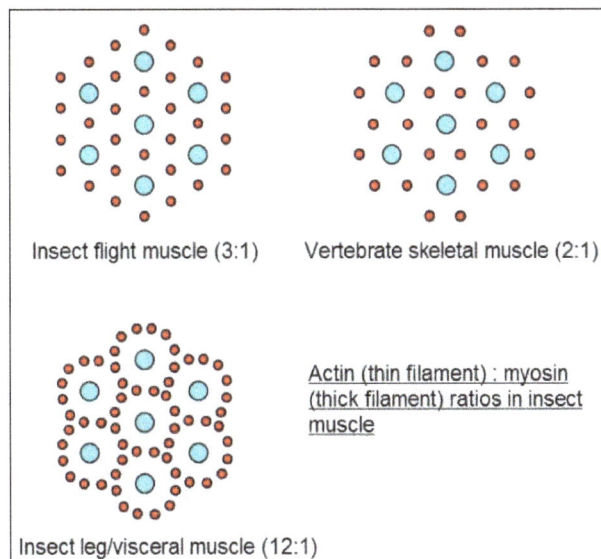

Insect flight muscle (3:1)   Vertebrate skeletal muscle (2:1)

Actin (thin filament) : myosin (thick filament) ratios in insect muscle

Insect leg/visceral muscle (12:1)

It is indeed very informative, from an engineering perspective amongst others, to compare insects with vertebrates. Both are highly evolved and advanced animals. Insects should no t be under-estimated. Insects often have different and novel solutions to life's problems.

Climbing in Insects

the insect foot is a remarkable mechanical device. The five segments of the tarsus can bend to take the insects weight, and the pretarsus has special equipment for gripping different surfaces. The pretarsus has a pair of claws (pretarsal claws) and between these is typically a retractible cushion called the arolium. The claws are used to grip a rough surface, whilst the arolium, aided by an oily secretion, is used to grip smooth surfaces. This allows most insects to walk up walls and across ceilings, or up plant stems and underneath leaves. Consider ants, they can support up to 100 times their body weight before their feet detach.

The mechanisms by which the arolium grips very smooth surfaces is not entirely understood. The secretion doe snot seem to act like simple glue. Clearly intermolecular forces must be involved. The figure below illustrates the anatomy of the ant pretarsus, externally in frontal view and internally in side-view:

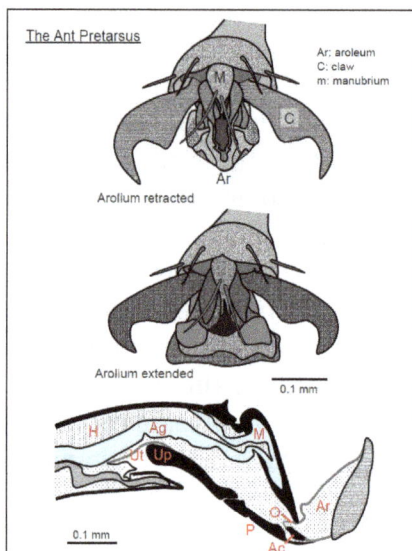

Iongitudinal section through pretarsus. Ac: arcus sclerite, Ag: arolium gland, Ar: arolium, H: haemocoel, M: manubrium, O: opening, P: planta, Up: unguitractor plate; Ut: unguitractor tendon.

The pretarsus articulates with the 5th tarsomere (5th tarsus segment). It consists of several cuticular plates (sclerites), the pair of pretarsal claws and arolium (Ar). The manubrium (M) is the topmost (dorsal) sclerite and and externally the planta (P) behind which is the unguitractor plate (Up). The unguitractor plate is operated by muscles via a tendon, the unguitractor tendon (Ut).

When the insect takes a step the following sequence of events occur:

1. The claws contact the surface and retract.

2. If the do not grip the surface (i.e. the surface is too smooth) then they retract fully and the arolium unfolds and extends, exposing the manubrium.

3. The arolium contacts the surface and adheres, assisted by glandular secretion.

4. The claws extend as the arolium deflates and detaches and the foot lifts off the surface.

The whole sequence of claw retraction and arolium deployment is triggered by muscular contraction pulling on the unguitractor tendon. The sequence of these events is as follows:

1. The unguitractor tendon is pulled, retracting the claws and pulling the whole pretarsus back into the 5th tarsomere.

2. The pretarsus is hinged dorsally with the 5th tarsomere and rotates as it is pulled back, planting the pretarsus on the surface.

3. The foot is fully rotated and further pulling on the unguitractor tendon now brings the unguitractor plate level to the surface, which pushes the arcus and planta upwards. This causes the arcus to rotate around the manubrium (about a horizontal axis) and the manubrium pushes down on the two side-arms of the U-shaped arcus. This pushing on the arcus unfolds the arolium downwards.

4. The arolium rotates downwards and pressure between the pretarsus and the surface expands the elastic arcus sideways. Thus the arcus converts a contact pressure into a sideways force which pushes the arolium out sideways. In some species of ants the dominant mechanisms for extending the arolium is different, instead as the pretarsus rotates fully about the 5th tarsomere, the pressure of the tarsomere pushing on the ventral surface of the pretarsus at the joint elevates hydrostatic pressure and pressure in the glandular secretions that fill the arolium. Thus, in this case hydraulic pressure expands the arolium sideways.

Insect larvae can also climb, but sometimes by very different mechanisms. The abdominal prolegs of caterpillars may be equipped with microscopic hooks (proleg crotchets), or they may be modified into suckers. Some caterpillars have a special escape mechanism: if disturbed they suddenly drop down from the leaf on which they were feeding by a silk thread (life-line) secreted by spinnerets in the head. At least two mechanisms of climbing back up this silk thread have been described. One mechanism involves the silk being wound-up between the third pair of thoracic legs, strung between them in figures of eight. The legs on one side will gather slack in the thread as the larva twists to one side, then pass the slackened thread onto the front leg of the opposite side. This foreleg then passes the thread to the middle leg of the same side, which then passes it onto the hindleg. The pattern repeats with the larva twisting to the alternate side - winding the silk about first one hindleg then the other.

Endocrine Glands in Insects

In insects, neurosecretory cells are numerous and have important functions. Recent work has revealed that there are two types of neurosecretory cells: type A which stain with para-aldehyde fuchsin, and type B which do not. All cells possess electron dense granules. The neurosecretory cells can produce blue colour from the reflected light due to the light scattering effect of colloidal-sized particles.

Protocerebrum

It is the most complex part of the insect brain consisting of several distinct cell masses and regions of neuropile. The pars intercerebralis, part of protocerebrum is located in the dorsal median region above the proto-cerebral bridge and central body. It contains two groups of neurosecretory cells that transport their secretions to the corpus cardiacum.

Sub-esophageal and Ventral Chain Ganglia

Neurosecretory cells are also found to be located in sub-esophageal and ventral chain ganglia. The sub-esophageal ganglia are composed of three fused ganglia that innervate sense organs, and muscles associated with mouth parts, salivary glands and neck region.

Corpora Cardiaca

The corpora cardiaca arise from the nervous system and are situated behind the brain in close association with the dorsal aorta. They receive axons from the neurosecretory cells in the brain and serves as storage and release sites for their secretions. That is why it also acts as neuro- haemal organ.

Four cellular elements have been recognised in this organ:

1.   The bulbous endings of neurosecretory axons whose perikarya are located in the dorsum of the brain,

2.   The perikarya of neurosecretory cells that send axons into nerves that supply various peripheral organs,

3.   Glia-like cells and

4.   Intrinsic corpus cardiacum cells.

Although, the corpora cardiaca are storage release centres, there are evidences that their own cells are capable of producing secretions.

Endocrine Organs of Epithelial Origin

Corpora Allata

Aggregation of ectodermal cells proliferated from the surface epithelium in the vicinity of the mouth parts are seen in the posterior margin of the brain. These cells form the gland, corpora allata. These glands are commonly paired and laterally placed (Periplaneta) or they may fuse to form a single structure (Rhodnius).

In the butterfly Pieris, few large cells with polymorphic nuclei are found in corpora allata. The corpus allatus of bug (Pentatoma rufipes) consists of almost syncytial mass of small cells with cell boundaries barely distinguishable. In phasmid (Bacillus rossii), a nearly columnar epithelium surrounding the embryonic lumen persists in the adult phasmids.

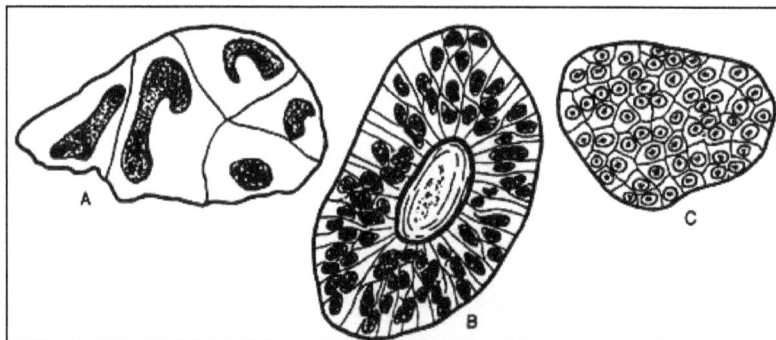

Histological structure of corpora allata of different insects : A. Pieris brasccicae, few large cells with polymorphic nuclei, B. Pentatoma rufipes, an alomost syncytial mass of small cells, and C. Bacillus rossii, a nearby column-nar epithelium surrounding the embryonic lumen

Thoracic Gland

These glands are found only in immature insects, with the exception of the Apterygotes. These are also called prothoracic glands or ecdysial glands. These glands consist of irregular masses of tissue of ectodermal origin that are usually intimately associated with tracheae.

The glands may or may not be innervated. Depending upon its final location, these glands are also identified as peri-tracheal or ventral glands. The cells of these glands show cyclic secretory activity, reaching a maximum between moults. The glands atrophy in the adults.

Ring Gland

The larvae of higher Diptera (true flies) contain a small ring of tissue, supported by tracheae, called the ring gland or Weismann's ring. The different cells that compose it are considered to be homologous with corpora allata, corpora cardiaca and the thoracic glands.

Some other Endocrine Glands

In Aeschnia cyanea, two types of endocrine cells were found to be located in midgut. One type is filled with dense granules and the other includes vesicles with an excentric core or has a loose filamentous appearance. These cells discharge their contents into the internal medium at the level of the basement membrane.

Neurohormones of the Brain

* Ecdysiotropin: Protocerebrum secretes ecdysiotropin or prothoracicotropic hormone (PTTH) or brain hormone (BH) that acts on ecdysial glands.

* Bursicon (Tanning hormone): Neurosecretory cells within the brain produce a blood borne hormone which triggers the tanning or darkening of adult cuticle. It is a protein hormone with a molecular weight of about 40,000. It is released into the fused thoracoabdominal ganglia. Bursicon is released after the emerged fly has dug its way out of the soil, but in some insects this hormone may be secreted slightly before or during the loss of the old skin.

* Eclosion hormone: Median neurosecretory cells of the brain produce eclosion hormone, which collects in the corpora cardiaca and is released into the blood at the time of switchover from pupal to adult stage. In most insects it acts upon neurons within the abdominal ganglia to initiate the pre-eclosion behaviour.

Hormones of Corpora Allata

* Juvenile hormone: Juvenile hormone is produced by the corpora allata. Chemically the hormone is methyl 10-epoxy-7-ethyl-3, ll-dimethyl-2, 6-tridecadienoate in moth Hyalophora cecropia. It is likely that considerable species variability may be found in chemistry of this hormone since the corpora allata vary in functional importance.

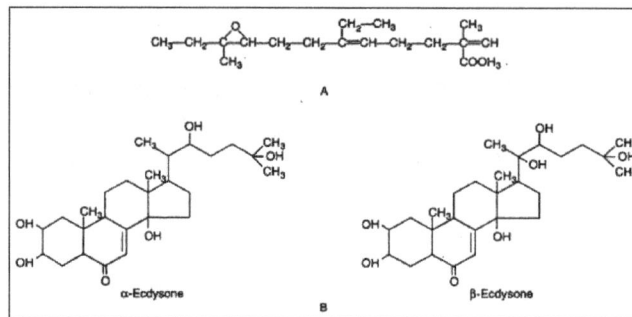

A Structure of Juvenile hormone of the month H. ceoropia B. Structures of ecdysones

Hormones of Ecdysial Gland

- Moulting hormone (Ecdysone): Ecdysial gland secretes the moulting hormone. It is a steroid, called ecdysone. Two types of ecdysones can be synthesised by insects – α-ecdysone and β-ecdysone. The latter steroid differs from α-ecdysone only by having a hydroxyl group at position 20. Similar β-ecdysone is also produced by crustacean arthropods.

- Functions of endocrine organs in insects: Insect hormones and neurohormones have been studied with respect to their involvement in a number of general physiological functions. Specifically, hormones and neurohormones influence development, diapause, mating and oviposition, metabolism, development of nervous system, control of circadian rhythms, regulation of dormancy, pheromone production and regulation of migratory behaviour.

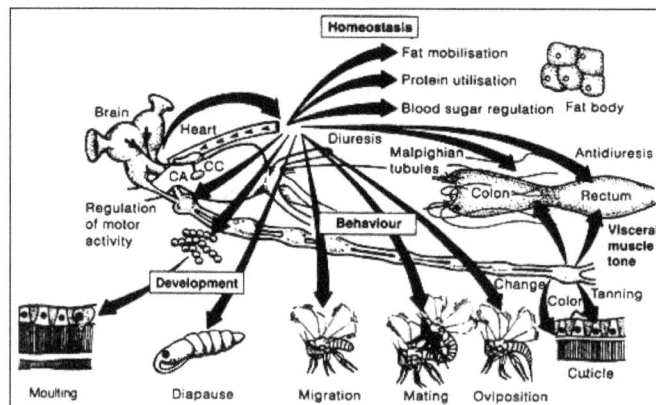

Major physiological functions regulated by neurohormones in insects

Endocrine Control of Growth and Metamorphosis

Upon emergence from the egg, the immature insects gradually increase in size to reach adults through some mechanisms called moulting. Moulting involves the periodic digestion of old cuticle, secretion of new cuticle (usually with larger surface area than the older one) and shedding of undigested old cuticle.

The last step – shedding of undigested old cuticle- is commonly referred to as ecdysis. Sometimes the term moult is also used synonymously with ecdysis. A typical insect goes through a series of moults, increasing in size with each step. Each developmental stage of the insect itself is called an instar, and the interval of time passed in that instar is referred to as stadium.

The whole developmental process by which the first instar immature stage of an insect is transformed into the adult insect is called metamorphosis. Metamorphosis can occur slowly in some insects or abruptly in others. The insects are divided into groups based on the type of metamorphosis.

During metamorphosis apterygote insects undergo very minor changes in form, as the immature instars differ from the adult only in size and development of gonads and external genitalia. These insects are called ametabola and undergo ametabolous development. Members of pterygota can be divided into two groups exopterygota where wings develop externally and endopterygota where wings develop inside the body of larvae.

The metamorphosis of exopterygotes is referred to as hemimetabolous or simple or incomplete development where the immature instars, commonly referred to as nymph, resembles the adults in many respects, including the presence of compound eyes, but they lack functional wings, gonads and external genitalia.

The remaining group of pterygota, those classified as endopterygota, undergo the developmental process referred to as holometabolous development or complete metamorphosis.

In these insects, the immature instars are referred to as larvae, which are quite dissimilar to the adults and are adapted to different environmental situations. The larvae typically lack compound eyes, may have mandibulate mouthparts and may or may not have thoracic or abdominal legs. In these insects, the changes in the transformation from the last instar larva to the adult are compressed into a single intervening instar, called pupa.

In most endopterygotes, the larval instars resemble one another except for a few minor morphological details and in size. However, some holometabolous insects pass through one or more larval instars that are totally different from the others (e.g., some members of Coleoptera, Diptera). This is called hyper-metamorphosis.

Endocrine Control of Growth and Metamorphosis

1. Hormones Required:

Brain hormone, Ecdysone, Juvenile hormone.

2. Tissues Involved:

Lateral neurosecretory cells of the brain, corpus cardiacum, corpora allata, prothoracic gland.

3. Hormonal Involvement:

Protocerebrum secretes brain hormone (BH) or prothoracicotropic hormone (PTTH) or ecdysiotropin which accumulate in the carpora allata and subsequently released into the haemolymph (except in Lepidoptera, in other insects BH is stored in corpora cardiaca). Through the haemolymph, PTTH reach to prothoracic gland and stimulate its secretory activity.

Secretion of PTTH from brain requires different stimuli in different insects. For example, in blood-sucking bug Rhodnius, abdominal distension resulting from feeding as well as a nutritional component of the diet stimulates secretion of PTTH from the brain. In Locusta, stretch receptors on the wall of pharynx seem to be involved whereas in Lepidoptera, the increasing weight of larva was thought to be the cue.

The prothoracic glands secrete α-ecdysone or moulting hormone (MH) which through haemolymph reach the target (epidermis) and is converted into active hormone 20-hydroxyecdysone, which initiates the growth and moulting activities of the cells.

Ecdysone fovours the development of adult structures and favours the moulting processes that terminate into successive larval instars. During pupal stage, ecdysone is needed for differentiation of the adult structures and the final pupal moult.

The corpora allata secrete juvenile hormone (JH), which promote larval development and inhibit development of adult characteristics. It has been described to have a "status quo" effect, which implies that it is more of an inhibitory substance than an active growth factor (Williams).

During the early larval instars, both MH and JH are produced prior to moult. In fact, JH interacts with MH to stimulate larval maturation during each stage of development. The concentration of JH evidently decreases toward the end of a larval instar, allowing the ecdysone to cause moulting.

The total picture here should be one of balanced interaction-synergism—between these two hormones to induce normal growth and differentiation, rather than a simple antagonism.

During the last immature instar, two separate and distinct peaks of ecdysone are present in both the holometabola and hemimetabola. The first one is low and in absence of JH, the epidermal cells are reprogrammed from larval to pupal commitment in holometabolous insects and from nymphal to adult stage in hemimetabolous insects.

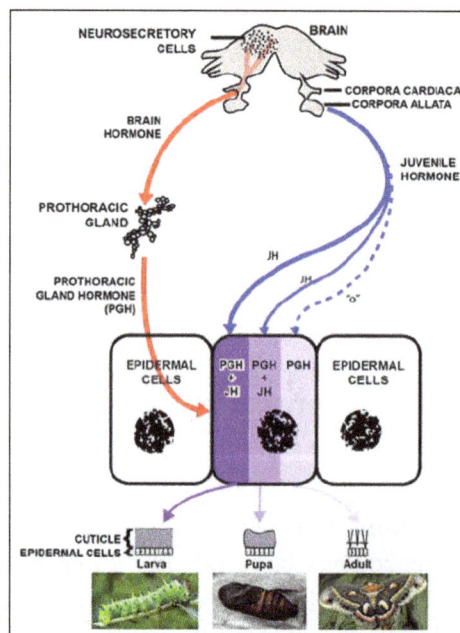

Hormonal control of development and metamorphosis in insect

The second ecdysone peak being high causes moulting (i.e. pupa in holometabolous form) in association with JH. The presence of JH at this stage prevents the imaginal discs (which have been growing slowly in the larval stages) from developing into adult structures during the pupal stage. Finally, the next high peak of ecdysone in the absence of JH causes adult development.

In hemimetabolous insects, there is a gradual progression toward adult stage, but there are also two ecdysone peaks prior to adult moult. In holometabolous form, on the other hand, the changes from immature to adult are compressed into the pupal stage. Thus, other than the pupal stage, there are no significant differences in the basic physiological processes controlling growth and metamorphosis in either of these two groups.

Molecular Mechanism

Ecdysone acts on several genes through receptor mediated processes and as a result, influences the synthesis of a number of enzymes as well as structural proteins. It has been shown that in Chironomus larvae, ecdysone produces puffing patterns in chromosomes identical to those patterns that occur during pupation. The puffing areas are the regions of newly synthesised mRNA. Ecdysone is thought to activate a gene that directs the synthesis of key enzyme of sclerotisation process.

In Drosophila, ecdysone was shown to cause two puffing patterns. First, it combines with ecdysone receptor, which in turn, causes early puff patterns. These early puff produces proteins that activate late puff patterns and also inhibit protein synthesis of the early puffs. At the same time, the late puff proteins are initially inhibited by the ecdysone.

The mode of action of JH is not known clearly, but is suggested that the hormone may activate or repress target gene or genes by binding with a specific receptor on a regulatory sequence of the DNA. It also acts by altering the number of ecdysone receptors present in a target tissue and JH may also influence the rate of target gene transcription in both larvae and adults.

Role of other Hormones in Metamorphosis

Eclosion

Eclosion hormone or EH is released from brain by a circadian clock and declining ecdysteroid titers. If ecdysone titer is artificially kept high, the release of eclosion and its activity are inhibited. This hormone influences many aspects of pupal-adult ecdysis, including the behaviour associated with ecdysis and subsequent degeneration of abdominal inter-segmental muscles used in the act of ecdysis.

Isolation of eclosion hormone gene from different insects suggested that EH hormone as well as the mechanism by which ecdysis behaviour is triggered is conserved among insects.

Ecdysis Triggering Hormone

It is the most recent hormone discovered that plays an important role in ecdysis. This 26 amino acid peptide hormone is synthesised by the epitracheal glands that are located segmentally in larvae, pupae and adults of Manduca sexta. According to Zitnan, this hormone may act upstream from the eclosion hormone in a series of cascade events leading to ecdysis.

Bursicon (Tanning Hormone)

Bursicon, commonly found in neurohaemal organs associated with the ventral chain ganglia is suggested to stimulate tanning and sclerotisation of the cuticle following ecdysis.

Hormonal Control of Reproduction

Like other higher multicellular organisms, reproduction in insects is a camplex process. Different stages of reproduction, starting from the production of male and female gametes to oviposition, are seem to be influenced by several hormones.

1. Spermatogenesis:

   Ecdysone controls the permeability of the testis walls to the humoral factor differentiating the spermatocytes. Juvenile hormone is shown to have some inhibitory effects on spermatogenesis in many insects.

2. Vitellogenesis:

   Vetellogenesis or egg yolk synthesis is also known to depend on JH from the corpora allata. In mosquitoes, juvenile hormone is required for egg development only during the early previtellogenic stages of development of the follicles. It promotes the growth of existing primary follicles up to the "resting stage". The allatohibin (20-HE) from the ovary has two important effects.

   It promotes early growth of secondary follicles leading to their separation as identifiable follicles, and is also important in initiating and sustaining synthesis and release from the fat body of vitellogenin. The third hormone necessary for ovarian follicular development is egg development neurosecretory hormone (EDNH). This peptide from the brain, controls ecdysone synthesis by ovary.

3. Oogenesis:

   The complex process of oogenesis differs among species. However, role of brain neurosecretory cells, ecdysone, corpora allata and ovary has been postulated in general. Hormones from corpora allata help in egg maturation through the incorporation of yolk into the oocyte.

   Some substances like farnesol and farnesyl methyl ether can mimic the effects of allata hormone and have a gonadotropic action. The secretions from the median neurosecretory cells of brain stimulate corpus allatum and protein synthesis (required for yolk formation).

   In addition to secretions from brain cells and corpora allata, ecdysone has been found to be involved in control of oogenesis in female mosquitoes. Following a blood meal, lateral neurosecretory cells secrete egg development neurosecretory hormone, which in turn, induces the ovary to secrete ecdysone. Ecdysone, in turn triggers the synthesis of yolk protein vitellogenin in the fat bodies. Juvenile hormones secreted by corpora allata also activate fat body and ovaries.

4. Fertilization:

   In many insects studied, ovulation (the passage of egg from the ovary into the oviduct) and oviposition, (passage of fertilized eggs to the outside, are closely linked. Both these events are affected by some peptides secreted by male accessory glands and neurosecretory products of brain. In Rhodnius, ovulation is controlled by a myotropic peptide originating in 10 identified neurosecretory cells of the pars intercerebralis.

The process of reproduction involves both the nervous and endocrine systems. The major centres are the neurosecretory cells of brain and the major events are the secretion of juvenile hormone by corpora allata, and either ecdysone production by ecdysial gland in immature insects or ecdysone biosynthesis by the ovary in adult insects. Both hormones act either independently or together in association with nervous system to make reproduction success.

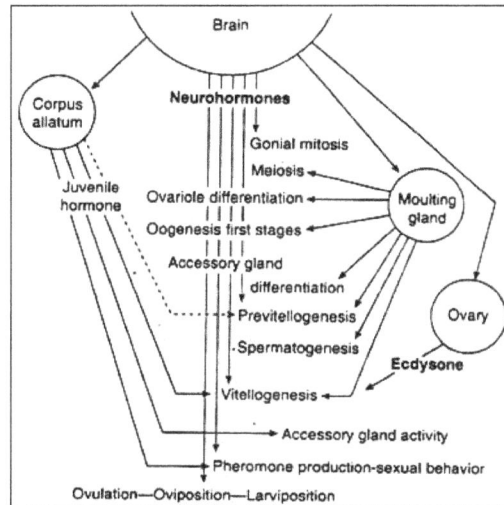

Hormonal control of reproductive events in insect

Hormonal Control of General Body Function

Digestion and Nutrition

In some insects movements of alimentary canal that complement the action of digestive enzymes and absorption of food is reported to be controlled partly by neurosecretory cells. In cockroaches, proctolin, a neurohormone, regulates peristalsis of hindgut. In Diptera, production and secretion of saliva are under control of a neurohormone (supposed to be 5-hydroxytryptamine, 5HT).

Circulation

Eclosion is known to control the eclosion heart-beat. At the time of adult eclosion from the pupal case and cocoon, the eclosion hormone influences cardio acceleration, which lasts only for 75 minutes of the eclosion period.

Excretion

The production of urine in insects is controlled by diuretic or antidiuretic hormones that are synthesised by pars intercerebralis of brain, corpus cardiacum, ventral chain ganglia and sub-esophageal ganglia. In Locusta, an antidiuretic hormone and a chloride transport-stimulating hormone have been shown to regulate both ion and water balance. In Periplaneta, proctolin is responsible for the hindgut motility which also acts as a neuromodulator helping in the excretion process.

Hormonal Control of Metabolism

Lipid Metabolism

In most insects, lipid remains stored in the fat body, an organ often suggested to be analogous to mammalian liver. Many hormones in different insects are known to stimulate lipid content of haemolymph. The adipokinetic hormone or AKH released from corpora cardiaca and prawn red pigment concentrating hormone (PRCH), an octapeptide have considerable adipokinetic activity. Interestingly, in cockroaches, AKH has no adipokinetic effect. In Locusta, storage of lobe of corpora cardiaca produces a hypoliparmic factor that opposes the action of AKH.

Carbohydrate Metabolism

In mosquitoes, median neurosecretory cell hormone (MNCH) suppresses glycogen synthesis. Diapause hormone is suggested to stimulate glycogen synthesis during diapause. Hormones secreted by corpora cardiaca inhibit oxidation of glucose in many insects.

Protein Metabolism

In insects hormonal control of protein metabolism has not been studied in detail (except for the tanning factors). However, it is reasonable to assume that some hormones may directly or indirectly influence the metabolism of amino acids in general.

Hormonal Control of Behaviour

Hormones play a very important role in insect behaviour. They either act as modifiers or as releasers of behaviour. As a modifier, a hormone alters the responsiveness of an organism. For example, the influence of ecdysone on the biting behaviour of female Anopheles, which does not take blood meals once ovarian development, has been initiated.

Ecdysone secreted by ovaries, inhibits biting when a hormone acts as a releaser, a specific behavioural pattern is exhibited within a few minutes of the secretion of that hormone. For example, ecdysis-triggering hormone and eclosion hormone in the moth Antheraea, act both as releasers in triggering the pre-eclosion behaviour and as modifiers in 'turning on' adult behaviour.

Migration

Migration in some insects occurs on the basis of endocrine changes correlated with particular environmental effects, such as crowding, food deficiency, and short day. For example, in aphids, crowding results in the production of winged forms (alate) instead of wingless, non-migrant (apterous) forms.

This is associated with the activity of corpus allatum. It has also been suggested that migratory behaviours depend on a particular balance between ecdysone and juvenile hormone in the haemolymph.

Dormancy

In response to the changes in the abiotic environment, insect development and general activity may undergo two kinds of suppression or dormancy. In response to adverse environmental condition, insects may slowdown metabolism and development; when conditions are no longer adverse, both functions resume immediately. This type of dormancy is referred to as quiescence.

Insects may enter into a state of metabolic and developmental arrest in response to certain environmental conditions which may or may not be adverse, but serve as indicators of the imminent onset of adverse condition. Development does not necessarily resume immediately with the return of favourable conditions. This type of dormancy is called diapause.

Diapause is controlled by hormones. The activity of the prothoracic gland is required for the termination of post embryonic diapause in silk moth, Platysamia. In other moths like Bombyx and Phalaenoides, a secretion from the sub-esophageal ganglia of the females induces diapause in the embryos. The arrest of development in larval and pupal diapause has classically been ascribed to an arrest of PTTH secretion.

In some species, the biological clock directly regulates PTTH secretion by a mechanism located entirely within the brain while in other species PTTH secretion may be restrained by high levels of JH. Indeed, high JH levels during larval diapause lead to the invocation of JH in regulation of PTTH release in non-diapausing moults. In some parasitic insects, development is closely attuned to the hormonal changes in the host. In these insects, diapause is induced in synchrony with diapause in the host.

The Respiratory System and Respiration

The basic insect respiratory system consists of a series of rigid tubes, called tracheae (singular trachea), connected to the outside via pairs of pores called spiracles (typically one pair per segment on the sides of the thorax and abdomen, lacking on certain segments). Air enters the system via the spiracles and the tracheae are air-filled. The spiracles can often be opened and closed and lead into short tracheae that enter a pair of longitudinal tracheal trunks, which are the main tracheal tubes. From these lateral tracheae branch smaller tracheae that supply the tissues with air. This supply is especially rich in the more active tissues, such as muscles, nervous tissues and the gut.

Tracheae also extend into the wings, running inside the wing veins. The tracheae branch until they reach a diameter of 2 to 5 micrometres (2-5 thousands of a millimetre) and then often enter stellate tracheole cells (transition cells) from which they emerge as finer branches called tracheoles, with diameters less than one micrometre. These tracheoles terminate inside the tissues, almost always as open-ended or blind-ended tubes about 200 nanometres in diameter (200 millionths of a millimetre).

Moth Bombyx mori. Bottom left: a trachea from the same insect. Note the spiral ridges (taenidia, singular taenidium) lining the inside of the trachea, this ridge is formed of cuticle and prevents the trachea collapsing. These ridges may be rings (annular) or spirals.

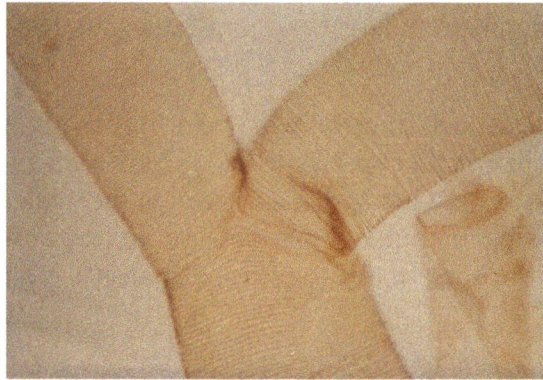

A branch in a trachea from Bombyx mori.

The outside cuticle of the insect extends inwards through the spiracles to line the inside of the main tracheae and in the smaller tracheae this cuticle is reduced to a thin membrane lining the lumen of the tracheae and forming the taenidia. The tracheoles are often said to lack this tough inner lining, but end. Apart from this lining, the walls of the tracheae consist of a single layer of flattened epithelial cells.

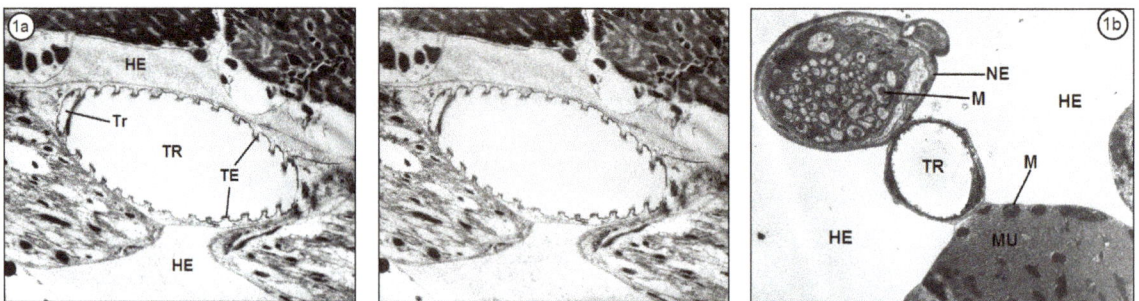

A cross-section through a trachea (TR) in the antenna of the rove beetle Aleochara bilineata, illustrating the taenidia (TE) and a small tracheole (Tr) branching off. The trachea is supplying nerves and muscles and is bathed in haemolymph (HE). A tracheole (TR) in the antenna of Aleochara bilineata, supplying a muscle (MU) and a nerve (NE). M: mitochondrion.

The tracheoles may terminate on the surface of a cell, such as a muscle cell, or they may penetrate inside the cell, either part way or even forming an extensive network inside and also covering the outside of the muscle. The supply is generally greater to flight muscles, especially of the fibrillar type.

The Role of Fluid in the Tracheoles

When an insect is at rest the ends of the tracheoles are filled with fluid. Textbooks sometimes state that this fluid is needed to dissolve the oxygen. However, oxygen diffuses faster in air than it does in water, and the fluid is actually a barrier to oxygen diffusion. Thus, when an insect exercises the fluid gets 'sucked' into the muscle cell until oxygen reaches the ends of the tracheoles. (This happens, at least in part, as the concentration of solutes build up in the exercising muscle, drawing in the water by osmosis). The fluid is there at rest because the tissues are bathed in fluid, though it may serve to reduce water-loss and dehydration through the tracheal system. When an insect hatches from its egg, its tracheal system is initially filled with fluid, but this fluid is actively absorbed by the tracheole cells until the system fills with air.

The Transport of Air through the Tracheal System

In the basic system described so far, air simply diffuses in through the spiracles and along the tracheae. Diffusion is rapid over a millimetre or even a centimetre, but is very slow at greater distances. This diffusion-driven system occurs in some small or not very active insects. In large and active insects, such as moths, butterflies, bees and wasps, diffusion alone is insufficient. These latter insects show breathing movements - that is they actively pump air through the tracheal system. This is why the abdomen pulses in these insects. Sometimes only the tergum of each abdominal segment moves up and down, as in beetles, or both the sternum and tergum, as in flies, or the side-walls (pleura) may be very flexible and also move in and out, greatly changing the internal volume of the abdomen, a sin moths and butterflies. In this way a rapid stream of air flows through the tracheal system. Certain spiracles may be used to take air in, others to expel air, e.g. air may be drawn in through the thorax and expelled through the abdomen. However, these circuits are not hard and fast and occasionally the direction of flow may be reversed.

Air sacs may facilitate this movement of air through the tracheae. Air sacs can occur in almost any part of the system, and in rigid structures like the head and thorax they may be permanently expanded, acting as reservoirs of air, whilst in the abdomen they may greatly inflate and contract (flatten and empty). The diagrams below illustrate the air sacs in the abdomen of a honey-bee worker. You may have noticed how the abdomen of a honey-bee pulsates as the muscles of the abdomen expand and contract the abdominal segments to fill and deflate the air sacs. If you have ever chased a bee or wasp, then you may also have noticed that the abdomen pulsates harder and faster with exertion. These breathing movements are intimately connected to the circulatory system of the insect also.

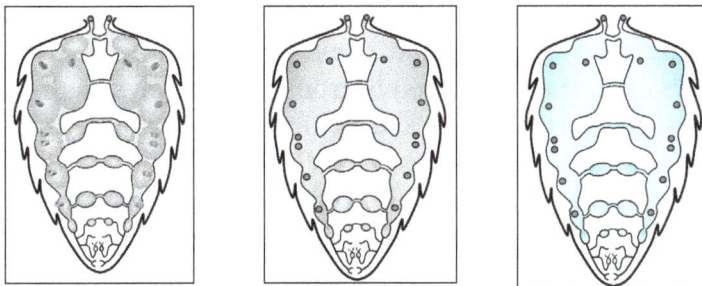

Air sacs in the abdomen of a honey-bee worker. The tubes (or circles) indicate the main tracheae entering/leaving the air sacs.

Control of Spiracle Opening and Closing

Insects lose most of their water through the spiracles. Being small they do not have much water to lose! It is not surprising, therefore, that insects typically only open their spiracles when they need more oxygen. An increase in carbon dioxide, a product of respiration, causes the spiracles to open more. Starvation and reduction in metabolic rate causes them to opening at the end of the trachea, opening at the end of the trachea, rather than the outermost opening of the rather than the outermost opening of the that is the internal opening of the atrium, that is the internal opening of the atrium, atrium (a naming convention you can personally find unhelpful since in insects lacking atria, the spiracle opens directly to the outside and in diagrams the external openings are generally labeled as 'spiracles' regardless and so you will be find myself referring to the most external opening, or even the whole structure, as a spiracle, atrium or not).The external opening of the atrium is often screened by spines, meshes and similar structures may cover the atrial openings in some insects, especially those from dry conditions, though both the spines and the atrium may trap water moisture, they may actually be more important in preventing dust from entering the tracheae. An internal spiracular valve often occurs between the spiracle and atrium, which can open or close the spiracle.

Temperature is an important factor. At low temperatures, when metabolism may be low, the spiracles are largely closed but open occasionally. At higher temperatures they may open and close periodically, and at still higher temperatures they may open continuously.

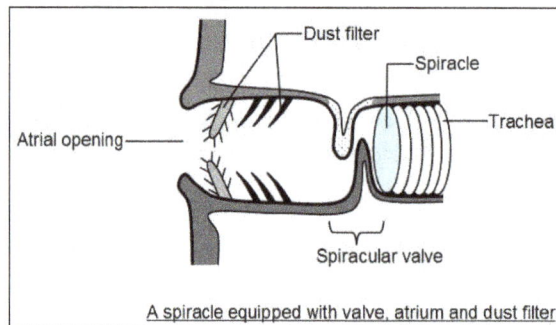

A spiracle equipped with valve, atrium and dust filter

Respiration in Aquatic Insects

Insects have to be able to obtain oxygen if they are to survive when submerged under water. Some insects have what is called a plastron mechanism - the hairs on their bodies are specially modified to trap a film or bubble of air when they dive (these hairs are designed to resist collapse under high pressures). In Dytiscus (Great Diving Beetle) and Notonecta (Water Boatman) air is trapped beneath the elytra (wing covers) and this air communicates directly with the insect's spiracles (openings to its airways or tracheae). These plastrons and sub-elytral air-spaces can function as what are called 'physical gills' - trapped pockets of air that can actually absorb more oxygen from the surrounding water. Oxygen diffuses across an air/water interface some three times faster than nitrogen, so as the insect takes up the oxygen from the trapped air bubble, lowering the partial pressure of oxygen and raising that of nitrogen within the bubble, oxygen diffuses in (when the partial pressure of oxygen in the water exceeds that in the air space) faster than nitrogen diffuses out. This maintains the air bubble for longer and in experiments in which the air was replaced by pure oxygen, the insect actually had to resurface for fresh oxygen sooner than when air was used, since with oxygen the partial pressure of oxygen in the bubble is always greater than in the surrounding water and although the insect takes more oxygen with it to begin with, it is unable to extract any from the surrounding water. In Notonecta (the Water Boatman) the hindlegs are used to drive water currents over the physical gills to irrigate them with fresh oxygenated water. Eventually all the nitrogen in the air space dissolves and then the insect must resurface to replenish its supply of air. In Dytiscus and Notonecta the trapped air volume can be regulated when under water, acting to regulate buoyancy.

Structures that have a higher affinity for air than for water, like plastron hairs, are called hydrofuge structures. Hydrofuge hairs also exist on the siphons of mosquito larvae. Some aquatic insect larvae, like mosquitoes, breath through a siphon - a tube with a spiracle at its tip that leads straight into the tracheal airways inside the insect. The spiracles are surrounded by hydrofuge hairs which repel water and prevent it from entering the system and drowning the insect. The water trough in the neighbouring meadow has dozens of these larvae that wriggle for cover whenever one's shadow passes over them. The hydrofuge hairs cannot repel oil, which is used to control mosquitoes by applying a film of oil to the water's surface; when the larvae try to breathe the oil enters through the spiracle and they drown. Eristalis, a type of hoverfly, has aquatic larvae that are called rat-tailed maggots because of the long tail-like extensible breathing siphon.

Some aquatic insects have a closed tracheal system which does not open to the outside air via spiracles. These include the dragonfly nymphs which have 6 double rows of lamellae lining the rectum and forming the branchial basket. Water is pumped over these tracheal gills, mostly in and out through the anus, and oxygen diffuses across to the trachea which fill the gills. Some aquatic insects have spiracular gills, like Simulium (black fly) pupae.

Bloodworms are red worm-like aquatic insect larvae that are often found at the bottom of ponds. These are the larvae of midges, like Chironomus, and certain flies (belonging to the true-flies or Diptera). They are red because they contain a form of haemoglobin (which contains 2 haem groups per molecule instead of 4 as are found in vertebrate haemoglobin) in their haemolymph ('blood plasma'). This haemoglobin is used to store oxygen, enough for two days in low oxygen conditions (as might occur in warm stagnant water). When oxygen is adequate they return to tracheal respiration and replenish their oxygen reserves. The larva of Chironomus has a closed tracheal system and absorbs oxygen across its 'skin' (cuticle) and also has so-called 'blood gills'. These are regions of the body wall that are very thin and project from the body surface as blood-filled sacs that are more-or-less devoid of tracheae. However, these do not seem to have a normal respiratory function, but may assist in recovery from oxygen starvation. The larvae of some mosquitoes have long anal papillae (projections) filled with tracheae and which are held in a current of water created by mouth brushes. However, the respiratory role of such tracheal-filled appendages is hard to determine experimentally and a normal respiratory function is doubted, though they may serve to excrete carbon dioxide.

Aquatic larvae may have closed tracheal systems, with no spiracles, or open tracheal systems, connected to the outside via spiracles. In the former, oxygen is absorbed through the cuticle, into the tracheal system (cuticular respiration). This may be facilitated by tracheal gills - structures filled with highly-branched, often feathery, tracheae that absorb oxygen across the cuticle. Often these gills are folds in the wall of the rectum (rectal gills) in which case the rectum may actively pump water across them, or they may be external appendages filled with tracheae. Sometimes these appendages are very long and bizarre looking, but have sometimes been shown to have no gill function, as their removal may have no effect on oxygen uptake, in which case they may function only in emergencies, or as an oxygen store, or they possibly act as flotation devices. In those larvae with functional spiracles, often at one end of the animal, the fluid is only actively removed from the system once the insect reaches the surface and starts taking in air, often through a snorkel-like siphon bearing the spiracles, as in mosquitoes.

Cuticular respiration is often still significant in terrestrial insects. Some butterflies absorb a significant proportion of their oxygen through the large surface area of their membranous wings. Generally, however, little oxygen is absorbed across the skin of insects.

The Circulatory System

Insects have an open circulatory system. This means that the internal organs and tissues are bathed in hemolymph, which is propelled actively to all internal surfaces by specialized pumps, pressure pulses, and body movements and is directed by vessels, tubes, and diaphragms. Without such constant bathing, tissues would die. The internal organs and tissues depend on the circulatory system for the delivery of nutrients, and to carry away excretion products, and as the pathway by which hormone messengers coordinate development and other processes.

Gas exchange in insects occurs via the tracheal system, which supplies all internal organs with tracheole tubules from spiracular openings in the body wall of terrestrial insects or from gill structures in aquatic insects.

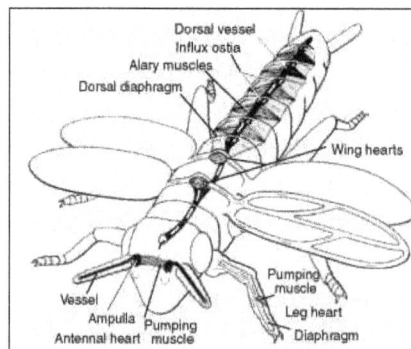

Delivery of the hemolymph to all tissues is so vital that a number of structures have evolved to ensure complete circulation. Principal circulatory organ is the dorsal vessel which is supported by the underlying dorsal diaphragm and the ventral diaphragm (omitted in the diagram). Circulation in the appendages is effectuated by accessory pulsatile organs (modified after Pass 2000).

However, the hemolymph has the capacity to dissolve carbon dioxide gas in the form of bicarbonate ions. A few insects live in low oxygen environments and have a type of hemoglobin that binds oxygen at very low partial pressures, but for the most part oxygen is supplied and carbon dioxide is removed by ventilation through the tracheal system. Besides the functions already mentioned, the circulatory system provides a medium in which battles are fought between the insect host and a myriad of invading disease microorganisms, including viruses, bacteria, fungi, and insect parasites. Principal participants in these interactions are the blood cells or hemocytes.

While maintaining the body tissues, the circulatory system is the medium in which homeostasis is ensured, including the regulation of pH and inorganic ions, as well as the maintenance of proper levels of amino acids, proteins, nucleic acids, carbohydrates, and lipids. Any change in the hemolymph quickly affects all organs bathed. The time for complete mixing of the hemolymph depends on the size of the insect, but it can be up to 5 min in a resting adult cockroach weighing about a gram. Any substance injected into a healthy insect will eventually appear at the extreme ends of all appendages in a few minutes, emphasizing the efficiency of the delivery mechanisms, which can be marvels of microhydraulic engineering.

Dorsal Vessel

The principal organ of hemolymph propulsion is the dorsal vessel, or at least it is the most visible organ specialized in hemolymph movement in insects. It forms a hollow tube which runs along the midline for the whole length of the body. Contraction of its circular musculature results in a contractile stroke (called systole), whereas elastic connective tissue strands, which connect the dorsal vessel to the body wall, are responsible for dilation and opening of the vessel (called diastole).

The dorsal vessel of most insects is not uniform in its course through the body and shows a differentiation into two regions which is reflected by their traditional denomination: the posterior part in the abdomen is referred to as the "heart," whereas the anterior part in the thorax and the head is the "aorta." Both terms are borrowed from better-known vertebrate structures and give an inaccurate impression of the different roles of those structures in insects. In phylogenetically ancestral insects, the flow of hemolymph in the dorsal vessel is bidirectional: upon contraction hemolymph flows toward the head in the anterior part and simultaneously toward the rear end in the posterior part of the dorsal vessel. The posterior directed flow is caused by an intracardiac valve in the abdomen and supplies the long caudal appendages (cerci and terminal filum), for example in some apterygotes and mayflies. However, in most insects the hemolymph flow is unidirectional in that the contraction of the dorsal vessel begins at the posterior end and advances forward as a peristaltic wave. In the pupae and adults of Coleoptera, Diptera, and Lepidoptera there is a regular change in the direction of the contraction wave, termed heartbeat reversal. Contraction waves of the dorsal vessel toward the head alternate with waves toward the rear of the body; between these contractions are short phases of rest.

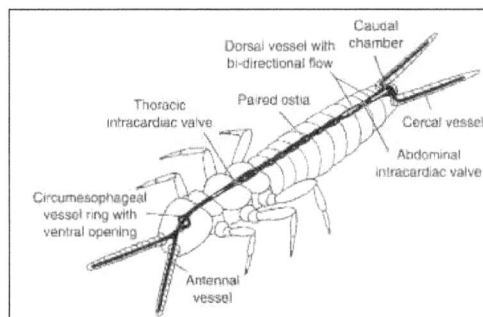

In the primitive insect Campodea (Diplura) the dorsal vessel exhibits
a bidirectional flow. This enables the supply of the antennae and the cercal
appendages by vessels connected to the dorsal vessel.

Hemolymph flows into the dorsal vessel through a pair of lateral openings (called ostia) in each abdominal segment. Valves which project into the vessel lumen close the ostia at the systolic contraction. Hemolymph emerges from the dorsal vessel at the open anterior end, and in insects with heartbeat reversal it usually also flows out the open posterior end. In Lepidoptera, there are two-way ostia which permit influx during forward phases and outflow during backward phases. The anterior end of the dorsal vessel opens just beneath or in front of the brain. This arrangement ensures a constant supply of nutrients and removal of waste products to and from the brain mass. In addition, the dorsal vessel is often intimately associated with the retrocerebral nervous system (including the hypocerebral ganglion, corpora cardiaca, and corpora allata complex) just behind the brain, which delivers neu-rohormones and possibly other hormones into the aorta at specialized release sites. Some insects have in addition outflow openings which are paired or unpaired and located more ventrally. In cockroaches, man-tids, and some orthopterans, the outflow openings are outfitted with sphincter-like valves and associated with segmental vessels laterally diverging from the heart. These vessels are formations of connective tissue with no inherent musculature, thus providing a simple channel to ensure lateral perfusion of the pericardial sinus.

The dorsal vessel is comprised of two rows of opposing pairs of muscle cells (collectively called myocardium). The cells of each row are offset, so that they cause a spiral-like peristaltic wave of contraction. The myocardium in all insects is spontaneously active, usually beginning in the embryonic stages. This type of heart control is termed myogenic, because the electrical activity underlying contractions arises in the myocardium itself. This is in contrast to a neuro-genic heart control present in crustaceans, such as crabs and lobsters, in which a barrage of nervous impulses drives the heartbeat from a discrete cardiac ganglion center. In insects, the myogenic pacemaker may be neurally or hormonally modulated. The basal heartbeat rate of most insects is around 60 beats min-1 at room temperature and at rest (e.g., in the American cockroach, Periplaneta americana, and the locust, Locusta migratoria). In adult house flies (Musca domes-tica), however, it ranges between 300 or more beats per minute during flight to zero beats for a while when at rest. The central nervous system of the adult house fly is composed of the brain and a thoracic ganglion mass, but lacks abdominal ganglia. Because of this unusual anatomy, the dorsal vessel of the abdomen can be separated from central nervous system input by experimentally severing the thorax from the abdomen. After this operation, the heartbeat of the fly becomes quite regular at around 60 beats min- 1. This indicates that the heart of the house fly is innervated by both inhibitory and excitatory motor neurons from the central nervous system.

Dorsal Diaphragm

The dorsal diaphragm is a fenestrated membrane which separates the upper pericardial sinus from the lower perivisceral sinus. It consists of connective tissue with associated muscles which are called alary muscles because in some insects, for example cockroaches, they resemble wings projecting laterally from the dorsal vessel of each abdominal segment. Although sometimes mistakenly thought to play a key role in heartbeat, the alary muscles are more properly called muscles of the dorsal diaphragm. Whereas the myocardium is specialized to contract rapidly and constantly, the ultrastructure of the alary muscles indicates infrequent and slow contractions. The muscles associated with the dorsal diaphragm may likewise be arranged in a loose network, as in Lepidoptera, or arranged like a weave surrounding the dorsal vessel, as in some Diptera; the functional role of these muscles is difficult to determine.

Ventral Diaphragm

The ventral diaphragm plays a prominent role in perfusing the ventral nerve cord of insects. Nearly 40 years ago Glenn Richards surveyed the ventral diaphragms in insects and found that insects with a well-defined ventral nerve cord in the abdomen also have a well-developed ventral diaphragm. In contrast, insects with the ventral nerve cord condensed into a complex ganglion structure in the thorax invariably lack a ventral diaphragm. This correlation suggests that the role of the ventral diaphragm is inexorably tied to perfusion of the ventral nerve cord in the abdomen. When present, the ventral diaphragm loosely defines a perineural sinus below and the perivis-ceral sinus above containing the gut. In some insects, the ventral diaphragm is a strong muscular structure with a great deal of contractile activity. The activity of the ventral diaphragm is dictated by innerva-tion from the central nervous system. In some large flying insects, the ventral diaphragm assists in hemolymph flow during thermoreg-ulation by facilitating the removal of warm hemolymph from the hot thoracic muscles to the abdomen for cooling. The intimate association between the ventral diaphragm in insects and perfusion of the ventral nerve cord is strengthened by considering the structure in cockroaches that takes the place of a proper diaphragm. In these insects, four stripes of muscle, together called hyperneural muscle,are near the back of each of the abdominal ganglia, and contract slowly but not in a rhythmic order. The muscles are electrically inex-citable, which means that they do not contract myogenically, as the myocardium does, but instead are neurally driven by motor neurons located in the ventral ganglia. Thus each of the ventral nerve cords in cockroaches has its own muscle supply that pulls it back and forth along the midline of the abdomen upon demand thereby increasing the contact and mixing between the ganglia and the hemolymph.

Accessory Pulsatile Organs

The dorsal vessel and the two large diaphragms are responsible for pumping hemolymph through the main body cavity. However, these organs are incapable of achieving circulation in body appendages such as the antennae, legs, wings, and various long abdominal processes (e.g., cerci and ovipositors). To supply these appendages, insects rely on special, small circulatory pumps, known collectively as accessory pulsatile organs or accessory hearts. As a rule, accessory hearts are separate from the dorsal vessel and function autonomously. The pumping organ is generally located at the base of the appendage and is connected to vessels or special diaphragms which guide the flow of hemolymph through the appendage. Accessory pulsatile organs are present in an astounding array of functional constructions. They are evolutionary novelties of higher insects and are absent in the phylogenetically ancestral insects. Antennal circulatory organs are nearly universal in insects, lacking only in groups with extremely short antennae, such as fleas and lice. The ancestral state of antennal circulation in insects is antennal vessels connected to the dorsal vessel as in certain apterygotes. The connection between the two vessels was probably lost during the further evolution of insects and replaced by autonomous pulsatile organs at the base of the antennal vessels. The anatomy of these organs differs widely in various insects. The best investigated antennal heart in terms of morphology and physiology is that of the American cockroach. Remarkably, the antennal heart of the cockroach functions not only as a circulatory pump but also as a neurohemal organ: hormones released into the ampulla lumen are pumped into the antennae where they most likely modulate the sensitivity of the numerous sensilla.

In leg circulatory organs, the flow of hemolymph is guided by longitudinal diaphragms instead of vessels. A diaphragm of connective tissue divides the inner cavity of each leg into two channels permitting a counter-current flow. Some insects have pulsatile leg organs with muscular attachment to the diaphragm. When the muscle contracts, one channel is compressed forcing hemolymph toward the thorax; at the same time, the other channel expands drawing hemolymph into the leg. Other insects utilize changes in the volume of an elastic tracheal sac in the legs as the driving force for hemolymph exchange. A common misconception is that insect wings are dead cuticular structures. However, the veins of the wings are filled with hemolymph to maintain living tissues, such as nerves and tracheae. Circulation is achieved by pumping organs in the thorax, which in ancestral winged insects are ampullary enlargements of the dorsal vessel. The wing hearts of most holometabolan insects, however, are muscular diaphragms which are separate from the dorsal vessel and which are either paired or unpaired. Recently, a further vital function of the wing hearts was discovered in Drosophila, namely that they are essential for the proper maturation of wings. Toward the completion of the wing formation process, they suck epidermal cells out of the wings which are necessary before the two cuticular surfaces of each wing bond together. Flies lacking functional wing hearts never develop flight ability.

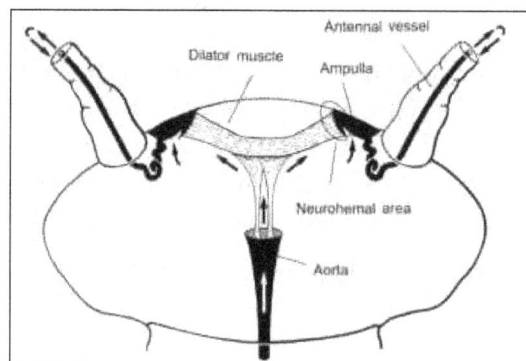

The antennal heart of the cockroach, Periplaneta americana, consists of a pulsatile ampulla at the base of each anten-nal vessel. The two ampullae expand by action of the interconnecting dilator muscle. Hemolymph rushes into each ampulla through a small ostium. When the muscle relaxes, the ampullae collapse by their own elasticity forcing the hemolymph into the antennal vessels. The hemolymph is conveyed to the end of each vessel and empties into the lumen of the antenna before returning to the head capsule; the two small muscles which extend to the anterior opening of the aorta are responsible for the suspension of the apparatus.

Finally it should be noted that accessory hearts may also partake in the hydraulic movements of body appendages. For example, the long sucking mouthparts of butterflies are uncoiled by action of special pumping organs in the head, and the lamellae of the antenna in scarabaeid beetles are spread out by action of the antennal heart.

Extracardiac Pulsations

Extracardiac pulsations of insects are the simultaneous contractions of intersegmental muscles, usually of the abdomen, that cause a sharp increase in the pressure in the insect body. The amount of movement accompanying each pulse is too small to be seen, but it can be readily measured as a slight shortening or telescoping of the abdomen as measured from its tip. The extracardiac pulses should not be confused with larger overt movements of the abdomen, especially in bees and bumble bees, that accompany ventilation during times or high activity or exertion such as flight.

Either the extracardiac pulsations occur in coordination with openings of certain of the spiracles, and therefore can play a role in ventilation, or they occur when all the spiracles are tightly closed, hence affecting hemolymph movement. The extracardiac pulsations become suspended only in quiescent stages of insect development, such as during diapause, but they can be evoked immediately upon disturbance or stimulation.

The extracardiac pulsations are driven by a part of the nervous system for which Karel Slama coined the name "coelopulse nervous system." The pressures induced by extracardiac pulsations are 100-500 times greater than pressures caused by contractions of the dorsal vessel and are transmitted by the hemolymph throughout the entire body of the insect, influencing hemolymph movement at some distance from the dorsal vessel and APO structures.

Tidal Flow of Hemolymph

A special condition of the circulatory system exists in some large, high-performance flyers, such as some Lepidoptera, Diptera, and Hymenoptera. To keep body weight at a minimum the amount of hemolymph is reduced and the volume replaced by large tracheal sacs. The body is often divided into two hemocoel compartments by an anatomical constriction between thorax and abdomen and may be additionally separated by a valve. The hemolymph in these insects is not circulated in the classical sense but is transported back and forth between thorax and abdomen by heartbeat reversal. The shift in hemolymph flow causes an alternating periodic increase and decrease in hemolymph volume in both compartments with a compensatory volume change in the tracheal system. This leads to ventilation especially of the large elastic tracheal sacs. Furthermore, the hemolymph in the wings oscillates in all veins simultaneously in correlation with heartbeat reversal and the resulting volume changes in the wing tracheae. Lutz Wasserthal called this periodic exchange of air and hemolymph in these insects "tidal flow" of hemolymph.

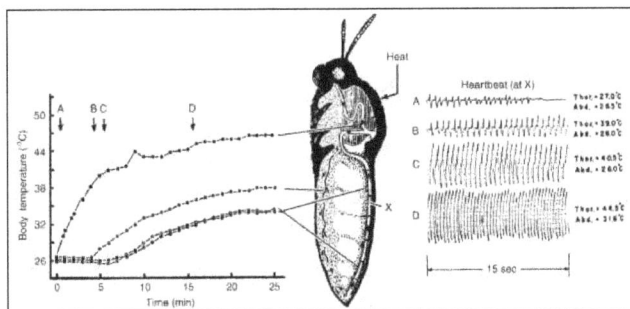

Control of thoracic temperature by central nervous control of dorsal vessel contractions during external heating of the thorax (heat). At the optimum temperature, hemolymph is pumped at maximum frequency and amplitude through the dorsal vessel to conduct heat from the thorax to the abdomen, where it is dissipated.

Thermoregulation

The use of the hemolymph in thermoregulation of flying insects was firstly described by Bernd Heinrich. The optimum temperature for flight muscle contraction in many insects, such the tobacco hornworm, Manduca sexta, is surprisingly high, up to 45 °C. Before this moth can fly, it must warm the thorax to near this temperature, which it accomplishes by means of a series of simultaneous isometric contractions of the antagonistic pairs of flight muscles that appear to the casual observer as "shivering," or vibrations of the wings.

A "thermometer" in the thoracic ganglia detects the proper temperature. When the thoracic temperature is below optimum, the central nervous system signals the dorsal vessel to circulate hemol-ymph slowly. When the thoracic temperature rises above optimum, the central nervous system brings about maximal amplitude and rate of heartbeat to drive hemolymph through the thoracic muscles. The increased hemolymph flow pulls heat away from the flight muscles in the thorax and eventually delivers hot hemolymph to the abdomen, where the heat is dissipated. Then relatively cool hemolymph is redelivered to the thoracic muscles by the dorsal vessel, completing the thermoregulation cycle. The warm hemolymph is then delivered to the head and percolates back past the ventral ganglia in the thorax to the abdomen, where the heat is dissipated. The cooler hemol-ymph is then delivered again to the thorax. The dorsal vessel and the very strong ventral diaphragm in the tobacco hornworm act together to move hemolymph. When the thorax is too warm, both the amplitude and the frequency of heartbeat contractions are increased, and the rate of delivery of hemolymph increases. When the thorax is too cool, amplitude and frequency of contraction of the dorsal vessel are decreased. The activity of the ventral diaphragm acts in concert with that of the dorsal vessel.

Thermoregulation of the flight muscles of the tobacco hornworm implies a sophisticated nervous control. The overall nervous control can be easily demonstrated by severing the ventral nerve cord between the thorax and abdomen. Then the moth can no longer thermoregu-late because the feed-back loop of temperature detection by the thoracic ganglia has been destroyed, and control over ventral diaphragm and dorsal vessel contractions has been lost.

Autonomic Nervous System

The tidal flow of hemolymph, the extracardiac pulsations, heartbeat reversal, and thermoreg-ulation all imply a very sophisticated control of circulation by the central nervous system. The central nervous system also plays a role in regulation of the respiratory system. The activities of circulatory and respiratory systems are coordinated by the central nervous system, perhaps to an extent not fully appreciated. It would be convenient and satisfying to be able to point out a particular part of the central nervous system and related peripheral nerves in insects that might comprise this regulatory system; however, beyond evidence that the meso- and/or metathoracic ganglia play a major role in some of these functions, entomologists know of no such discrete structure or structures, possibly because these interregulatory functions have been undertak-en by different parts of the nervous system in different insects. It is known that insects have a number of regulatory mechanisms that can be recruited to achieve such control, from motor and sensory neurons to neurosecretory neurons and neurohormonal organs located throughout the insect hemocoel.

The Excretory System

Excretion is the process whereby an organism eliminates metabolic wastes and unwanted chemicals from its system. Metabolism is the sum total of all the chemical reactions occurring in the cells and body. Some products of these metabolic reactions are toxic and so must be pro-cessed or eliminated from the body. Others are simply materials that are present in excess and so must be eliminated as waste. The process of excretion is quite different to defecation, which is the removal of undigested food wastes from the gut. However, the gut of many animals also has a role in excretion as some materials may be excreted into the gut and eliminated with the faeces. In insects most excretory products are excreted into the gut lumen and eliminated along with faecal matter. Excretion is also important in eliminating excess water and other unwanted chemicals that may be ingested and enter the body fluids, such as plant poisons and excess salts.

One of the main functions of excretion is to remove excess nitrogen. Nitrogen enters the diet in the form of amino acids, nucleic acids and certain salts. One of the main products of excretion in aquatic organisms is ammonia. Ammonia contains nitrogen and is a small molecule which dissolves readily in water. This allows it to be easily excreted into the surrounding water. However, this becomes a problem for terrestrial organisms. Ammonia is toxic to cells and so must be quickly ejected from the body, however, being water-soluble it is typically ejected in solution, which requires water. The mammalian solution is to convert the ammonia into a less toxic substance called urea. This conversion takes place in the liver: the ammonia produced by cells enters the bloodstream where the liver removes it, converts it into urea which again enters the bloodstream to be excreted by the kidneys. Being less toxic, the urea can be temporarily stored and excreted in a concentrated solution, requiring less water.

Birds and reptiles have a better water-conserving system; they excrete uric acid (or urate salts). Uric acid is not readily soluble in water and is of low toxicity and so can be excreted with very little water. The dry excreta of birds is a mixture of faecal matter and uric acid crystals and when water is scarce birds can produce very dry excreta.

Arthropods, including insects, have adopted similar solutions. Woodlice, which are not insects but crustaceans, are only partially adapted to terrestrial conditions, preferring moist habitats, but they do excrete ammonia. Interestingly they can vent off ammonia gas, rather than relying on the wastage of water to remove the ammonia in solution. Insects are better adapted to dry conditions, although aquatic insects and some insect larvae excrete ammonia, most terrestrial forms excrete uric acid (or salts of uric acid called urates, such as ammonium urate).

If one considers how small an insect is and how rapidly a small drop of water may evaporate, then one realises that insects have outstanding water-conserving systems. Bedbugs (Rhodnius) can survive for weeks without ingesting any water! Some insects can tolerate extremely dry conditions and may excrete uric acid as a dry crystalline powder, along with bone-dry faeces! Insects generally produce only trace amounts of urea.

Malpighian Tubules

The main excretory organ of the insect is the Malpighian tubule. Insects contain anything from 2 to 150 or more Malpighian tubules depending on the genus. Malpighian tubules are tubular outgrowths of the gut. They typically develop as pouches emerging from the junction between the midgut and the hindgut, though there actual final position varies - they may be attached to the midgut, hindgut or the midgut-hindgut junction as is the case with our ant above.

Each Malpighian tubule is a blind-ending tube whose lumen is continuous with the lumen of the gut. Each consists of a single layer of epithelial cells, forming the tubule wall, enclosed by an elastic membrane (basement membrane - a fibrous and porous protein mesh). In most insects there is a thin layer of striated muscle around this membrane. Typically muscle cells spiral around the distal end (the end furthest from the gut) of the tubule, causing it to twist and turn in gentle writhing movements as the muscles contract. The proximal end (near the gut) may be coated in circular and longitudinal muscle fibres, giving rise to peristalsis or squeezing movements which empty the contents of the tubule into the gut. In some cases, such as in caterpillars, the Malpighian tubules on each side (3 on each side in this case) empty into a small bladder, which then empties into the gut. In this case only the bladder may be muscular and its lumen is lined by cuticle (suggesting that the bladder is an extension of the hindgut).

The tubules do not just hang around in the air! The body cavity of the insect is filled with a fluid, usually colourless, called haemolymph. This fluid bathes the organs and tissues and is circulated around the insect body. The tubules are also typically loosely or firmly anchored in place by the tracheae which attach to them.

The twisting and turning of the Malpighain tubules presumably keeps them in contact with fresh haemolymph (perhaps by circulating the heamolymph around the tubule). Metabolic wastes and other unwanted chemicals that entered the insect system pass into the haemolymph, or are excreted into the haemolymph by the cells. These include nitrogenous waste and plant toxins such as alkaloids. It is the job of the Malpighian tubules to keep the haemolymph cleansed of these wastes - they remove wastes from the haemolymph and then excrete them into the gut lumen.

Outside the muscle layer is a 'peritoneal covering' of cells with embedded tracheoles, which carry oxygen to the Malpighian tubules which their mitochondria use to generate the needed ATP by aerobic respiration.

How do Malpighain Tubules work?

Waste materials and excess water pass from the haemolymph into the Malpighain tubules, by crossing the epithelial wall of these blind-ended tubes. Recent evidence shows that these cells contain pumps, proteins called proton-secreting V-ATPase. These proteins use energy in the form of ATP to pump protons into the lumen of the Malpighian tubule. Protons are positively charged and to maintain charge balance the removal of protons from the epithelial cells, into the tubule lumen, is balanced by the inward movement of potassium ions, which move from the haemolymph, into the epithelial cells and then out into the tubule lumen also. The diagram below shows a section through a segment of a Malpighian tubule. The epithelial cells have microvilli (fingerlike projections) projecting into the tubule lumen and are rich in mitochondria (green stripy rods) which produce the ATP required by the pumps. A model of how ion transport across the epithelium is thought to take place is illustrated.

The detailed structure of the cell at top right has been simplified to illustrate some of the transport mechanisms. The V-ATPase is shown as the orange circle pumping protons (H^+) into the tubule lumen.

Removal of the protons from the epithelial cell makes the cytoplasm more negatively charged and also sets up a concentration gradient (that is an electrochemical gradient is established) and this attracts positive ions, such as sodium (Na^+) and potassium (K^+) into the cell from the haemolymph.

The influx of these positive ions drags in negative chloride ions to balance the charge. These ions move across the cytoplasm of the cell, the so-called transcellular pathway. (see transport across membranes).

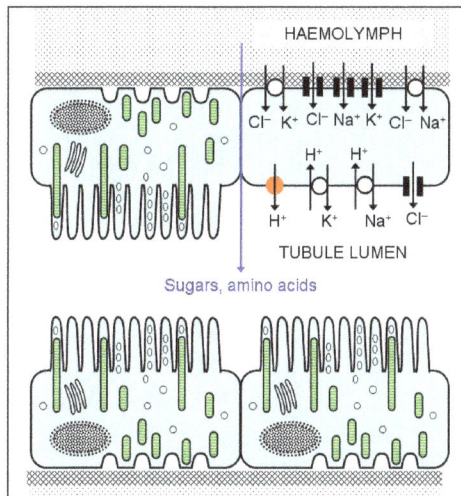

The flux of ions across the epithelial cell also draws across water, by osmosis. This probably takes place largely by the paracellular pathway, that is between the epithelial cells. Sugars and amino acids are swept along by the water into the tubule lumen. Since these materials are useful they will be reabsorbed later further downstream.

Other small molecules (small enough to cross the basement membrane) will also move into the tubule through this pathway. The transport of a substance which depends directly on ATP, such as the pumping of the protons in the Malpighian tubule, is called active transport. The transport of the other ions and water is passive (by facilitated diffusion) in of itself, but is dependent on proton transport and so indirectly dependent on ATP. This mode of transport is called secondary active transport, e.g. the transport of potassium.

In dry conditions many insects can produce a very concentrated urine, indeed one that is 'bone-dry'. However, many insects ingest large quantities of water when feeding, such as blood-sucking insects, and in this instance the rate of fluid-flow through the Malpighian tubules increases a thousandfold or more. Indeed, the rate of fluid transport in these tubules is said to be higher, gram for gram, than any other tissue. Two hormones, released into the haemolymph, can stimulate Malpighian tubules to rapidly increase their rate of fluid transport: 5HT (5-hydroxytryptamine) and a peptide hormone. Increased excretion is triggered by an increase in uric acid following a meal, which presumably triggers the release of the diuretic (urine-producing) hormones.

Of course, not all the fluid transported through the tubules is excreted. The proximal (basal or lower or downstream) sections of the tubules, along with the hindgut (especially the rectum) reabsorb some of the water, depending on need, and other useful substances, such as certain ions, sugars and amino acids, so as to produce a final urine of the 'desired' concentration. It is in this proximal or lower part of the tubule that uric acid is transported into the tubule, against a concentration gradient, and precipitates as crystals, e.g. of insoluble potassium urate as the urate combines with the high potassium content of the tubule lumen. In some insects these crystals can be seen filling the lumens of the proximal ends of the tubules. Presumably, peristalsis then moves these crystals along into the gut. Potassium and some of the chloride are recovered in this way, producing a urine high in sodium.

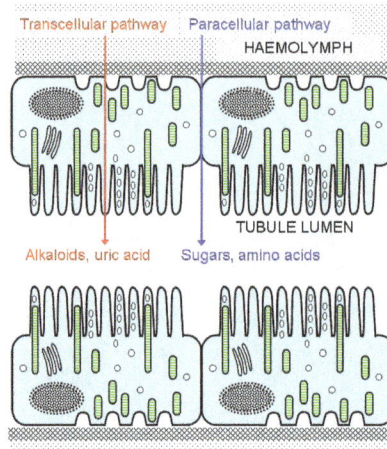

Some small organic molecules are also actively transported into the tubule lumen by the transcellular pathway, including alkaloids (plant compounds which may be toxic to the insect).

Uric acid, mostly in the form of negatively charged urate ions, is also actively transported by the transcellular pathway, though the exact mechanism is not well understood. This urate transport occurs in the proximal tubule and the urate combines with the potassium transported into the tubule to form insoluble potassium urate crystals. These crystals form roughly spherical concretions in the tubule lumen. The microvilli in the proximal tubule seem to undergo a cycle of elongation, as the urate concretions form, and retraction as the lumen fills up with urate waiting to be transported into the gut.

Once in the gut, remaining water may be reabsorbed as needed and the remaining urate excreted with the faeces, or separately. The midgut is divided from the hindgut by the pyloric sphincter and when this sphincter is closed the hindgut receives only the contents of the Malpighian tubules.

The mechanism of excretion demonstrated by the Malpighian tubule is one largely dependant on 'secretion' of unwanted materials, such as urate and excess sodium. This contrasts with the mammalian kidney which relies on ultrafiltration (filtration through microscopic pores), which removes most materials from the blood except large proteins and cells, followed by reabsorption of what the body needs to keep, such as sugars and amino acids. However, there is some filtration in the Malpighian tubule, namely the influx of materials through the paracellular pathway, having filtered across the basement membrane. Sugars and amino acids filtered in this way are then reabsorbed, as in the mammalian case. Similarly, there is some secretion in the mammalian kidney, for example the secretion of protons and ammonium in acid-base balance and the secretion of some drugs such as penicillin. However, the emphasis is different with the Malpighian tubule relying more on secretion, the mammalian kidney on filtration.

Other Mechanisms of Excretion

Some insects, such as the silverfish, springtails and aphids have no Malpighian tubules. Stick insects may have three types of Malpighian tubules. Clearly much remains to be learnt about excretion in insects. In addition to excretion by Malpighian tubules, insects often exhibit storage excretion in which waste materials are sequestered safely and kept inside special storage cells. For example, the fat body may contain urate cells which accumulate urate crystals throughout the life of the insect.

pH Regulation and other Functions of Malpighian Tubules

The main function of Malpighian tubules may be the elimination of nitrogenous waste, but hand in hand with this comes the task of water conservation (eliminating waste whilst conserving water when necessary) or osmoregulation - regulating water content of the insect body and also regulation of ion balance. Considering their involvement in cleansing body fluids of unwanted materials it is not surprising that excretory organs typically have major roles also in regulating acid-base balance. Enzymes only work within a narrow range of acidity or pH and so an organism has to excrete excess acid or excess base to maintain the correct pH of its body fluids. Malpighian tubules also have a role in acid-base balance. The V-ATPase actively excretes protons and hence excess acid (an acid is a chemical which generates protons in solution as the protons are the true source of acidity).

Calcium is also excreted in large quantities by the Malpighian tubules of some insects. Generally, some of the Malpighian tubules, or one specific segment of the tubules, takes on this function. These tubules often become distended as they fill with calcium salt crystals. Some insects make use of this calcium in the construction of their burrows or larval cases, such as the helical calcium carbonate shells of some spittlebug (Ptyelus) larvae.

Finally, the Malpighian tubules of some insects may assume a glandular function in the secretion of silk.

The Reproductive System

The reproductive organs of insects are similar in structure and function to those of vertebrates: a male's testes produce sperm and a female's ovaries produce eggs (ova). Both types of gametes are haploid and unicellular, but eggs are usually much larger in volume than sperm.

Most (but not all) insect species are bisexual and biparental — meaning that one egg from a female and one sperm from a male fuse (syngamy) to produce a diploid zygote. There are, however, some species that are able to reproduce by parthenogenesis, a form of asexual reproduction in which new individuals develop from an unfertilized egg (virgin birth). Some of these species alternate between sexual and asexual reproduction (not all generations produce males), while others are exclusively parthenogenetic (no males ever occur).

Sexual reproduction might well be the most important "adaptation" ever acquired by living organisms. It provides a mechanism for shuffling and recombining genetic information from two parents to create new ("hybrid") genotypes that can be tested in the fire of natural selection. Only phenotypes that withstand the "heat" can participate in the next round of reproduction.

External vs. Internal Fertilization

As long as primitive arthropods lived in the water, their sperm could simply swim from the male's body to the female's body where fertilization could occur. But in order to adopt a terrestrial life-style, animals that engaged in such external fertilization had to protect their sperm from desiccation. The solution, still used today by myriapods and insects, was to encapsulate large numbers of sperm within a water-tight lipoprotein shell secreted by the male's accessory glands. These "packages" of sperm are known as spermatophores. In myriapods and primitive hexapods (e.g. Collembola), males leave spermatophores on the ground where they may be found and picked up by a passing female. Silverfish and bristletails have more elaborate courtship activities in which the male leads his mate to a freshly deposited spermatophore.

Stalked spermatophore of a collembolan

Today, all of the more "advanced" insects exhibit internal fertilization — males deposit their sperm inside a female's body during an act of copulation. This novel adaptation, which appeared soon after insects diverged from their myriapod-like ancestors, presumably ensured that more sperm found their way to a receptive female. But the genetic programming for spermatophore production still persists in most modern insects. After a male deposits his spermatophore inside a female's reproductive system, she digests the lipo-protein coat and uses it as a source of additional nutrition for her eggs. In some cases, the quality (or quantity) of this nuptial gift may even determine whether a female accepts or rejects the male's gametes.

Sex Determination

Like humans, most insects have a single pair of chromosomes that carry the genetic information for determining an individual's gender. If an embryo inherits a pair of "X" chromosomes, it will develop as a female; if it inherits one "X" and one "Y", it will develop as a male. The "XX" female is said to be homogametic; the "XY" male is heterogametic. In this case (as in humans) the male's contribution determines the offspring's gender. Some insect species have no "Y" chromosome at all — males have just one "X", and females have two. A similar condition is found in some parthenogenetic species of aphids in which "maleness" occurs through the loss (degeneration) of one chromosome during embryogenesis. In both cases, the males end up with an odd number of chromosomes (2n-1).

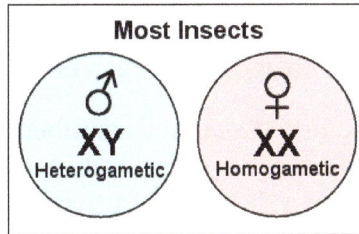

In Lepidoptera and Trichoptera, however, the homo- and heterogametic sexes are reversed: females are heterogametic and males are homogametic. To distinguish this system from standard X-Y sex determination, these sex chromosomes are designated "W" and "Z" (instead of "X" and "Y"). Thus, a female butterfly is "WZ" and a male butterfly is "WW". In this case, the female's contribution determines the offspring's gender.

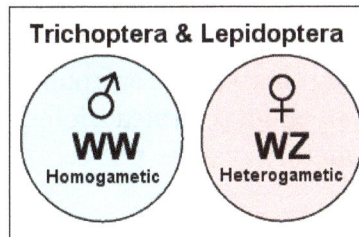

A third method of sex determination, called haplo-diploidy, is found in all Hymenoptera, many Thysanoptera, some scale insects (Hemiptera/Homoptera), and a few weevils (Coleoptera). These insects have diploid, homogametic females ("XX"), but all of the males are haploid — they develop by parthenogenesis (asexually) from unfertilized eggs. Primary oocytes undergo meiosis to form haploid eggs, but meiosis is unnecessary in primary spermatocytes because the cells are already haploid. Unmated females can lay eggs that will develop into males. Once a female mates and receives sperm from a male, she has two options:

1. She can produce a female offspring by opening the valve at the base of her spermatheca to release sperm onto the egg as it passes through her oviduct.

2. She can produce a male offspring by closing the spermathecal valve and preventing any sperm from reaching the egg.

Control over the gender of offspring has proven to be a useful adaptation for some insects. A biased sex ratio that favors females over males can reduce competition for limited food resources and increase the reproductive potential of the population. Bees, wasps, and ants form large colonies of queens and workers (all female) in which males are produced only sporadically as needed for reproduction.

Female Reproductive System

The main function of the female reproductive system are egg production and storage of male's spermatozoa until the eggs are ready to be fertilized. The basic components of the female system are paired ovaries, which empty their mature oocytes (eggs) via the calyces (Calyx) into the lateral oviduct which unite to form the common (median) oviduct. The gonopore (opening) of the common oviduct is usually concealed in an inflection of the body wall that typically forms a cavity, the genital chamber. This chamber serves as a capulatory pouch during mating and thus is often known as the bursa copulatrix. Its external opening is the vulva. In many insects the vulva is narrow and the genital chamber becomes an enclosed pouch or tube referred to as the Vagina.

Two types of ectodermal glands open into the genital chamber. The first is the spermatheca which stores spermatoza until they are needed for egg fertilization. The epermatheca is single and sac-like with a slender duct, and often has a diverticulum that forms a tubular spermathecal gland. The gland or glandular cells within the storage part of the spermatheca provide nourishment to the contained spermatozoa.

The second type of ectodermal gland, known collectively as accessory glands, opens more posteriorly in the genital chamber.

Each ovary is composed of a cluster of egg or ovarian tubes, the ovarioles, each consisting of a terminal filament, a germarium (in which mitosis gives rise to primary oocytes), a vitellarium (in which oocytes grow by deposition of yolk in a process known as vitellogenesis) and a pedicel. An ovariole contains a series of developing oocytes each surrounded by a layer of follicle cells forming an epithelium (the oocyte with its epithelium is termed a follicle), the youngest oocytes occur near the apical germarium and the most mature near the pedicel.

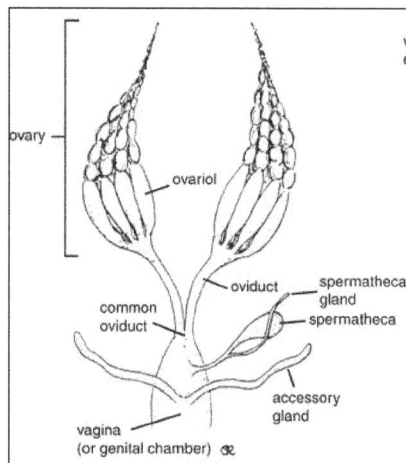

The different types of ovariole is based on the manner in which the oocytes are nourished.

- Paniostic ovariole: Lacks specialized nutritive cells so that it contains only a string of follicles, with the oocytes obtaining nutrients from the haemolymph via the follicular epithelium. e.g.Cockroach.

  Ovarioles of the other two contains trophocytes (nurse cells) that contribute to the nutrition of the developing oocytes.

- Telotrophic ovariole: (Acrotrophic) The trophocytes are confined to the germarium and remain connected to the oocytes by cytoplasmic strands as the oocytes move down the ovariole, e.g. bugs.

- Polytrophic ovariole: A number of trophocytes are connected to each oocyte and move down the ovariole with it, providing nutrients until depleted, thus individual oocytes alternate with groups of smaller trophocytes. e.g. moths and flies.

Accessory glands of the female reproductive tract are often called as colleterial or cement glands, because their secretions surround and protect the eggs or cement them to the substrate. e.g. egg case production in mantis, ootheca formation in cockroach, Venom production in bees.

Structure of Egg

- Chorion

- Vitelline membrane

- Micropyle

- Periplasm with yolk.

Male Reproductive System

The main functions of the male reproductive system are the production and storage of spermatozoa and their transport in a viable state to the reproductive tract of the female. Morphologically, the male tract consists of paired testes, each containing a series of testicular tubes or follicles (in which spermatozoa are produced) which open separately into the mesodermally derived sperm duct or Vas deferens which expands posteriorly to form a sperm storage organ or seminal vesicle. Tubular paired accessory glands are formed as diverticula of the vasa deferentia. Sometimes the vasa deferentia themselves are glandular and fulfil the functions of accessory glands. The paired vasa deferentia unite where they lead into the ectodermally derived ejaculatory duct (the tube that transports the semen or the sperm to the gonopore).

Accessory glands are 1-3 pair, either mesodermal of ectodermal in origin and associated with vasa deferentia or ejaculatory duct. Its function is to produce seminal fluid and spermatophores (sperm containing capsule).

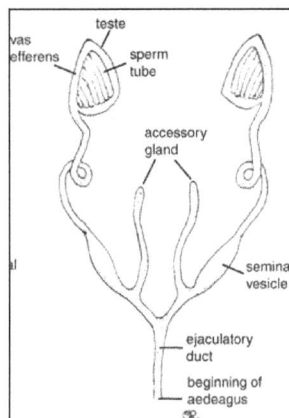

References

- Insect-nervous-systems, BioTech: cronodon.com, Retrieved 20 March, 2019

- Insect-locomotion, BioTech: cronodon.com, Retrieved 23, 2019

- Endocrine-glands-in-insects, insects: notesonzoology.com, Retrieved 2 February, 2019

- Insect-respiration, BioTech: cronodon.com, Retrieved 12 April, 2019

- Circulatory-system-insects, insects: what-when-how.com, Retrieved 17 January, 2019

- Insects-excretion, BioTech: cronodon.com, Retrieved 15 August, 2019

- Rreproductive-system: cals.ncsu.edu, Retrieved 5 May, 2019

Chapter 4

Sensory Systems and Behaviour of Insects

The physiological capacity of organisms that provides data for perception is known as sense. Insects are able to sense various kinds of stimuli such as light, heat, chemicals and mechanical pressure. The topics elaborated in this chapter will help in gaining a better perspective about the branches of sensory systems and behaviour in insects.

Insect Senses

Touch

The bodies of insects are covered with sensitive hairs. These are most numerous and most sensitive on parts such as the antennae (feelers) and on the lower sections of the legs.

Each hair is a stiff modified part of the cuticle and the underlying skin to which it is jointed. A sense cell beneath the joint is supplied with a nerve fiber and, when the hair is touched, the mechanical disturbance of the associated sense cell or the resulting chemical changes causes signals to pass along its nerve and the insect "feels".

If a person is blindfolded and asked to retrieve an object at the far side of a room he or she will grope about using their hands to "feel" their way across the room. In doing this he or she is acting very much in the way that an insect does: for the insect waves its antennae about, relying more on touch than we do, and less on sight.

Besides the sensitive hairs insects have other sense organs. These are modified dome-shaped patches of cuticle which, like hairs are associated with a sense cell and its nerve. They occur on the tail spines (cerci), the wings, and the lower parts of the limbs. Stresses in the cuticle move the dome, which disturbs the sense cell.

There appears to be a direct link between the control of flying and the touch receptors in the lower sections of the legs, because many insects commence flying as soon as the weight is taken off the limbs. Presumably these limb receptors are stimulated continuously while the insect is standing, and the succession of signals passing to the central nervous system overrules the nerves supplying the wings. The insect flies when these signals stop or when the higher centers overrule them.

Smell and Taste

The receptors for these senses are concerned with the detection of chemical substances and are called chemoreceptors. Because of this relationship between the two senses it is often difficult to say definitely which is used in a particular instance. A rough distinction is that smell detects chemicals which exist in the air as vapor, while taste recognizes liquids or solids which come into contact with the receptors.

Receptors of chemicals seem to occur on all parts of the body. All have one thing in common: they have a thin cuticle and are supplied with one or more bipolar nerve cells. Those for smell are concentrated on the palps and antennae in the honeybee.

The sense of smell is used for various purposes – sexual attraction, recognizing the odor of their own species, finding suitable sites in which to lay eggs, and finding food. The males of certain moths are attracted by the scent of virgin females, for example. Some moths are able to scent their mates at a distance of three kilometers (two miles). Certainly social insects are able to detect members of their own species, even of their own colony. Often an ant of the same species but from another colony will invade the home of a neighboring colony. It will be recognized as foreign because its different smell can be detected and so will be evicted or killed. Some ants will follow trails left by others and so they are guided on their food-foraging expeditions.

Female insects are often attracted to a suitable site in which to lay their eggs by its smell. Parasitic forms. Plant-eating insects are often attracted to the plants on which they feed by the odors that these emit. In this way night-flying moths are guided to such flowers as honeysuckle. Many drably-colored flowers attract flies because of their fetid smell. Such features play an important part in helping to ensure that pollination is carried out. It is interesting that the newly emerged females of an ichneumon fly find the scent of the pine unattractive and fly away from it. When they are sexually mature, however, the scent attracts them and they fly back to it to lay their eggs on the host insects that they parasitize. By offering substances which to us taste bitter, sweet, salt, or acid (sour) it is possible to study an insect's sense of taste. Bees are particularly fond of sweet substances such as some sugars, but on the other hand they will reject many sugars to which flies are attracted. Honeybees also find man-made sweetening substances such as saccharin distasteful.

The taste receptors occur mainly in the mouth and on the mouthparts, the antennae, and the lower parts of the legs of many insects. Insects can literally taste with their feet.

Temperature and Humidity

Often on a hot day one is aware that it is also muggy. Insects, too, are receptive to both humidity and temperature. It is probable that the receptors in the human body louse, Pediculus, are tufts of specialized hairs, and hairs of various kinds are thought to be humidity receptors. At present it is not known if the receptors are stimulated by changes in the suppleness of the cuticle or whether they are directly receptive to water vapor.

Very few heat receptors have been identified positively. Some blood-sucking insects (e.g., mosquitoes) are known to locate warm-blooded prey by detecting the heat that their bodies produce though the receptors have not been isolated.

Sight

Light visible to ourselves lies between red and violet, but though many insects are not so sensitive to the red end of the spectrum they can see well into the ultraviolet which we are unable to see.

Honeybees are even able to perceive the plane of vibration of polarized light, which faculty they are able to use as a means of direction-finding.

Insects respond to light in three ways: through the whole of the body surface which appears to be sensitive, through simple eyes called ocelli, and through compound eyes.

In many simple animals the body is receptive to light even though no obvious eyes are present. Simple eyes occur on the head near the compound eyes, usually in a group of three. They have no focusing mechanism. (Many insects have no simple eyes.) Simple eyes probably measure the light intensity, and in some insects, at any rate, prepare the compound eyes for receiving light, thus alerting the insect.

Compound eyes are the main organs of sight. The outer part of each eye consists of numerous units or facets varying in number from about ten to thirty thousand in different insects. They are usually hexagonal, forming a honeycomb-like mosaic. Each facet or lens is at the top of a cone-shaped tube (ommatidium) at the bottom of which are the light receptive cells. All the ommatidia together produce a mosaic of spots of light, each spot representing the part of the field of view in line with a particular ommatidium. In effect each picture consists of a series of dots – rather like a printed picture in a newspaper. The picture an insect sees, however, is indistinct. Insects can certainly distinguish shapes and can recognize certain patterns, but their eyes are best suited to pick out moving objects or those moving across their flight path. A dragonfly is an expert, for example, at catching its prey on the wing. In most insects the field of view of both compound eyes overlap so that they have stereoscopic vision somewhat like our own.

Many insects can distinguish colors. Karl von Frisch was a pioneer in experiments on this. He placed a dish of sugar on the blue-colored board of a colored strip. The bees came to the dish and fed. When the dish was removed the bees still went to the blue color where the dish had been. To show that they were in fact "seeing" the strips in color and not merely as different shades of gray he placed several dishes on a checkered board, each square having a different intensity of gray except one which was blue. The bees went for the blue color that they had learned to visit in the first place.

Hearing

This sense is not equally developed in all insects. In fact it is likely that some insects are deaf. The ranges of hearing of different insects cover a wider frequency band than that to which the human ear is sensitive.

Hearing organs are of two kinds, sensitive hairs and tympana. Few insects have the latter. The sensitive hairs are similar to those used for touch. Each tympanal organ consists of a thin membrane connected to sensitive cells that are supplied with nerve fibers. The positions of the tympana vary considerably. Crickets and long-horned grasshoppers have them on the fore-legs, but short-horned grasshoppers have them on the first segment of the abdomen. Others may have them on the thorax or the abdomen.

The purpose of the insect's hearing organs seem to be the location of the other sex for mating purposes. Crickets and the like chirp or stridulate, and researchers have succeeded in attracting unmated female crickets to a loudspeaker from which noises made by young males were coming.

Insect Mechanoreception

Mechanoreception is the sense that allows insects to detect their external and internal mechanical environments, including physical orientation, acceleration, vibration, sound, and displacement. The integument and internal organs contain a wide variety of mechanore-ceptors. Prominent receptors, such as surface hairs that mediate touch, or auditory organs, have been studied extensively, but many other physiological functions also depend on mechanosensory signals.

Arthropod mechanoreceptors are divided into two morphological groups: Type I, or cuticular, and Type II, or multipolar. Type I are ciliated receptors, associated with the cuticle, and have their nerve cell bodies in the periphery, close to the sensory endings. They can be subdivided into three major groups. Hair-like receptors are found on the outer surface in a variety of shapes and sizes, from long, thin hairs to short pegs and scales. A sensory neuron is closely apposed to the base of the hair, and its dendrite contains microtu-bules ending in a structure called the tubular body. It is assumed that movement of the hair compresses the ending, with the tubular body perhaps providing a rigid structure against which the compression can work. Hair receptors can contain additional sensory neurons, such as chemoreceptor neurons in taste hairs. Campaniform (bell-shaped) sensilla are also found on the outer surface, particularly in compact groups near the joints, where they detect stress in the cuticle. Stress moves the bell inward, compressing the dendritic tip containing the tubular body. Chordotonal receptors are generally found further beneath the integument, although they can be connected to the integument by attachment structures. They serve several functions, including hearing and joint movement detection. They often lack tubular bodies but all have dense scolopales surrounding the dendrites and often have multiple mechanosensory neurons.

Type II mechanoreceptors are nonciliated neurons, whose central cell bodies have many fine dendritic endings, each of which is apparently mechanosensitive but lacks the detailed structures seen in Type I receptors. Type II receptors are found in many Structures, predominantly associated with mesodermal tissues, including the musculature, where they detect muscle tension.

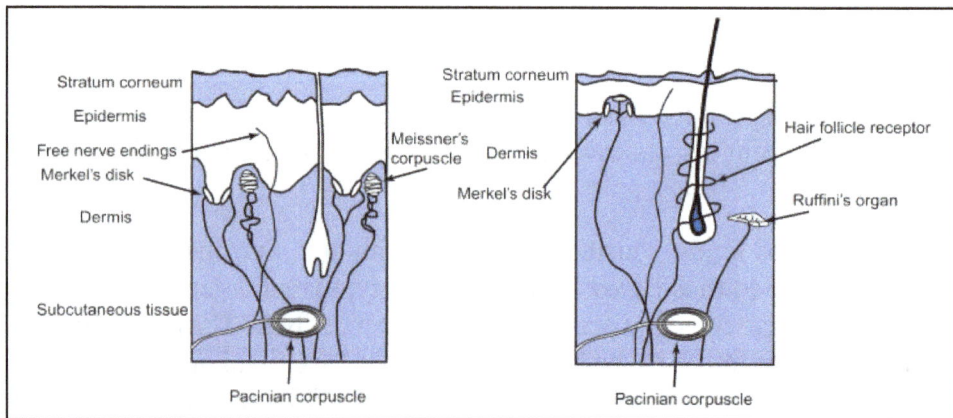

Studies of mechanoreceptor morphology have used many techniques, including light microscopy, scanning and transmission electron microscopy, and immunohistochemistry. Receptor electrophysiology has been studied by three basic methods: (1) Extracellular recordings observe the receptor currents flowing along the axon. (2) Epithelial recordings measure the current flowing through the relatively low resistance of the thin socket tissue or through a cut hair. (3) Intracellular recordings give direct measurements of membrane potentials and currents. Important evidence applicable to insect mechanoreceptors has been obtained from similar structures in arachnids and crustaceans.

Mechanosensation is commonly viewed as a three-stage process in which a mechanical event is first coupled to the receptor cell membrane by mechanical structures, then transduced into a receptor current at the cell membrane, and finally encoded into action potentials for transmission of information to the central nervous system (CNS).

Development of Mechanoreceptors

Type I sensory neurons are surrounded by specialized sheath cells of varying numbers and names, although the terms trichogen (hair-forming) and tormogen (sheath-forming) are commonly used for the innermost two layers of sheath cells. Development of these cells has been well characterized in several species, but especially in Drosophila external bristles, where many of the genes involved have been identified. A single sensory organ precursor cell divides to give two different secondary precursors, IIA and IIB. IIA divides to form one trichogen and one tormogen cell. IIB gives rise to the neuron, another sheath cell, and sometimes an additional glial cell. The neuron then forms an axon that grows into the CNS. A variety of other noncellular structures, including sheaths, are also found, particularly in dendritic regions. The development of Type II receptors is less well understood.

Mechanical Components

Extracellular tissues, often with elaborate structures, surround the sensory cells. These structures modify the spatial and temporal sensitivities of the receptors, and are often designed to interact with the outside environment or other parts of the animal, such as cercal hairs detecting air movements or hair plates detecting joint rotation. External structures usually allow detection of mechanical events at some distance from the sensory cell, but make the displacement at the receptor cell membrane smaller than the original movement. Estimates of this attenuation suggest that threshold movements at the cell membrane leading to sensation are in the range 1-5 nm.

Transduction and Encoding

Mechanically activated ion channels, probably located at the tips of the sensory dendrites, transduce the mechanical stimulus into a receptor current. These transducer channels are permeable to potassium ions, which are more concentrated in the receptor lymph space. The molecular identity of transducer channels is unknown, but there is evidence for both the transient receptor potential (TRP) and acid-sensitive (ASIC) channel families. Mechanotransduction currents are very sensitive to temperature, with activation energy values of 12-22kcalmol-1.

Current flowing through the channels causes a receptor potential that is encoded into action potentials using several different sodium and potassium currents. Action potentials propagate into the central nervous system along axons in nerve roots of the segmental ganglia. Afferent axons have a size range of 1-20 |im and conduction velocities are typically 1-5ms-1. Information is transmitted from mechanore-ceptor axons into the central nervous system via cholinergic synapses.

Central, Peripheral and Humoral

Modulation

Many mechanoreceptors receive GABAergic inhibitory efferent innervation close to the output synapses of their axon terminals. This presynaptic innervation modulates afferent mechanoreceptor information. Some mechanosensory neurons are also modulated by efferent innervation in the periphery and by circulating chemicals such as biogenic amines, of which octopamine has been most thoroughly studied. Octopamine directly excites spider mechanoreceptors in the periphery, but the evidence is less clear in insects.

The extent and functions of peripheral modulation in insects remain to be seen. It is the latest in a series of surprises about the complexity of insect mechanotransduction, but probably not the last.

Thermoreception in Insects

A thermoreceptor is a temperature (or heat-flow) sensor. Hygroreceptors are moisture-sensors or water-sensors. It is common to find dual-purpose thermo and hygroreceptor sensilla in insects - so-called thermohygroreceptors. These can take many different forms but are usually pointed/ trichoid hairs, or blunt pegs/cones which may be enclosed inside a pit (coeloconic sensilla). These sensilla are non-porous, except for a terminal moulting pore (which may lead to confusion with gustatory sensilla which have a pore in their tip to allow liquids to contact the sensory dendrites inside); some thermo-hygroreceptive sensilla also contain olfactory sensory cells, in which case the walls may be multi-porous, with the pores allowing diffusable gases to reach the sensory cells within.

A typical thermo-hygroreceptive sensillum contains 2-5 sensory cells (though three cells is the most common), with four striking features. The dendritic outer segments of only two of the cells extend into the lumen of the peg. The dendrite which does not enter the peg is usually flattened and forms characteristic lamellae (layered sheets). The outer dendrites which are found within the lumen of the peg, fill the lumen completely (there is no receptor lymph cavity) and tightly. 4) These latter dendrites usually contain many prominent microtubules.

Lamellated outer dendrites were first discovered in the cave beetle Speophyes lucidulus. Corbière-Tichané assumed a sensitivity to light, but since Speophyes is a cave beetle, to infrared (IR) radiation in particular. This was supported by the fact that the lamellae react with osmium, as do photoreceptor outer segments (due to the presence of rhodopsin). However, electrophysiology has revealed hygro- and thermoreceptive units in the antenna, which respond to changes in air temperature, but not to IR emission. Thus, the presence of a lamellated dendrite appears to correspond to thermoreceptor activity. Many other electrophysiological studies have confirmed that non-porous sensilla are thermohygrorecetors.

The two dendrites occupying the lumen, correspond to hygroreceptor units, identified electrophysiologically, generally one dry-air and one moist-air sensitive cell. The fact that these dendrites are well-packed with microtubules, often surrounded by electron-dense material, is suggestive of a mechanoreceptive role, as in the tubular body of other mechanoreceptors. It has been suggested that a swelling process, in the presence of moisture, could lead to the exertion of mechanical forces on the dendritic membranes of the hygroreceptors, which would then be specialised mechanoreceptors. Thermo-hygroreceptive sensilla apparently occur very infrequently on the insect integument, and so are probably often overlooked and are incompletely understood.

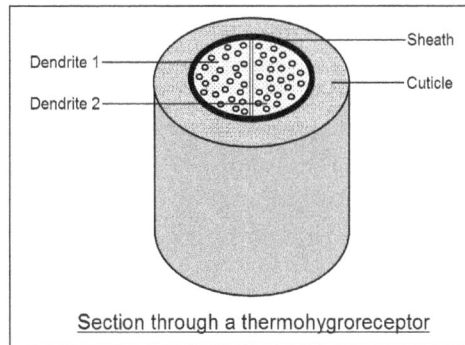

Section through a thermohygroreceptor

A section through a typical thermohygroreceptor peg.

Such a peg is often quite short and embedded in a pit, or dome with a central pore, in the insect cuticle, forming a coeloconic (peg in a pit) sensillum. The pit presumably holds moisture, increasing the sampling time for the hygroreceptor to function, or may protect the pegs, which do not have flexible sockets, from breakage. Two dendrites are tightly packed inside the cuticle, surrounded by a sheath which is an extension of the scolopale sheath. In addition, these sensilla typically have a thermoreceptive dendrite (e.g. a cold-sensitive receptor) in the base of the peg.

Such a way that the inner lymph-cavity shrinks and the wall constricts around the sensory dendrites. In this state the inner wall squeezes and places pressure on the dendrites, which get squeezed against the expanded inner wall. This squeezing of the dendrites may activate the dendrites to send signals to the sensory cell body and, if the stimulus is strong enough, the cell body will relay the signals to the central nervous system (as action potentials travelling in sensory axons). Thus, these moisture sensitive dendrites are thought to function as pressure sensors and so are fundamentally mechanoreceptive. Thus, this sensillum is a device called a hygromechanical transducer.

Of the two dendrites which enter the lumen of the peg, and branch, one is thought to be moisture-sensitive, the other dry-sensitive. It is the moisture-sensitive dendrite which responds to pressure as the inner wall expands and constricts around it. The dry receptor, responds when it expands as the wall dries, shrinks and moves outwards. The moisture-sensitive receptor also responds to increasing air pressure, the dry receptor to decreasing air pressure, supporting the hygromechanical transducer model. However, the exact mechanism of operation of these sensors is not firmly established.

Double-walled Olfactory Receptors

These are grooved pegs, often coeloconic. It is not always clear whether, or not, double-walled grooved pegs have pores in their walls, since these pores possess no pore tubules and may be clogged by a secretion. Electrophysiological studies reveal that many of these sensilla are dual-purpose olfactory and thermo- and/or hygroreceptors, although some are exclusively thermo-hygroreceptors, none have been found to have an exclusively olfactory function.

Double-walled Sensilla in Aleochara

In Aleochara bilineata, about 20 of these sensilla occur on the tip-most segment of the antenna (antennomere F9) increasing in density towards the tip. One such sensillum is shown in the photograph at the top of the page. They are blunt, slightly tapered pegs, about 5um in length, and less than 1um across at their base, and taper only slightly for most of their length, before tapering very steeply to a point in the last 0.5um of their length.

In cross-section these pegs have an inner and an outer cuticular wall, connected by cuticular spokes which are separated by outer lymph chambers. The outer surface of the pegs contains 9-13 grooves, the spokes connect to the outer wall where the grooves lie. Sometimes the grooves possess clear central channels, which are blocked by electron-dense material. (In electron micrographs, materials appearing dark are said to be electron dense or darkly-staining, since they absorb and scatter electrons from the electron beam). The chamber within the inner wall contains some lymph, but is largely occupied by closely-spaced dendritic branches. Initially a single dendrite enters the peg, then branches into three dendrites about mid-way along the peg. Each of these branches apparently forks again, as typically 4 or 5 branches are seen in sections towards the tip of the peg. The tapering tips of these pegs consist entirely of cuticle, with a diminishing number of external grooves.

Left: A cross-section through a double-walled peg-sensillum of Aleochara
bilineata of the type shown in figure 1 as seen in the transmission electron microscope.

The outer surface contains grooves running the length of the peg. The cuticular wall (dark grey) consists of an outer cylinder, the outer wall (ow), joined by radial spokes (s) to an inner cylinder, or inner wall (iw). Between the spokes chambers of darkly-staining liquid (lymph) can be seen - the outer lymph chambers (ol). In the centre is a cavity containing (lightly-staining) lymph (il, inner lymph chamber) and, in this case, five dendritic branches (d) can be seen in cross-section. Although it has been reported that there are no obvious pores in the outer wall (Skilbeck and Anderson, 1996) grooves filled with electron-dense material have been seen to connect the outer lymph chambers to pores in the outer cuticle wall. This topic has two apparent pores lying in the grooves, so although pores are hard to detect in these sensilla they do appear to be present. However, these pores seem to lack the usual pore tubule arrangement seen in the majority of chemoreceptors, but this is typical of double-walled olfactory receptors.

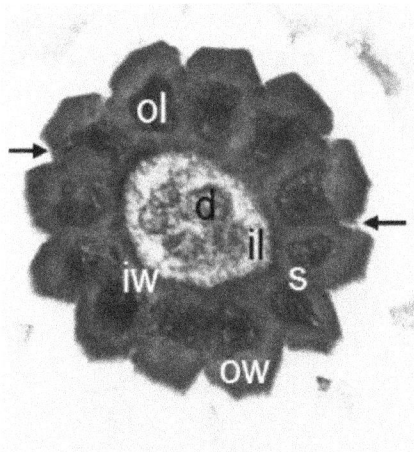

These pegs therefore seem to be what are called
double-walled multiporous sensilla.

Sensilla of this type, in other insect species, have been shown to have a thermoreceptive and an olfactory function, and this is generally their assumed function, though some are reported to be thermohygroreceptors only. In this type of sensor there is no apparent ultrastructural difference between the thermosensitive and chemoreceptive dendrites. None have a lamellated structure and all extend to the tip of the peg. It is not clear how these receptors detect temperature and moisture. Closer to the base of the peg, the spokes tend to disappear and the inner wall, apparently an extension of the sheath more tightly encircles the dendrites, but it is difficult to see how any of these sensilla could detect moisture by a hygromechanical mechanism.

A sensillum of this structure is characteristic of a dual-function olfactory-thermoreceptor, which might or might not have an additional hygroreceptive function. The actual function of this sensillum could be unequivocally observed by electrical recordings in the living specimen, but in insects determining sensory modality from structure gives a reliable indication of a sensor's function. Such measurements have been performed on similar sensilla in other insect species, which is how their function was first ellucidated.

Sometimes one or more of the spokes are absent, or only partially formed and not connecting to the outer wall. Thus the outer lymph chambers are apparently connected. In particular, the spokes disappear completely towards the base of the peg, followed by the grooves as the base of the peg becomes smooth (and indistinguishable from the bases of single-walled basiconic pegs).

Photoreceptors

Compound Eyes A pair of compound eyes are the principle visual organs of most insects; they are found in nearly all adults and in many immatures of ametabolous and hemimetabolous orders. As the name suggests, compound eyes are composed of many similar, closely-packed facets (called ommatidia) which are the structural and functional units of vision. The number of ommatidia varies considerably from species to species: some worker ants have fewer than six while some dragonflies may have more than 25,000.

Externally,each ommatidium is marked by a convex thickening of transparent cuticle, the corneal lens. Beneath the lens, there is often a crystalline cone secreted by a pair of semper cells. Together, the lens and the crystalline cone form a dioptric apparatus that refracts incoming light down into a receptor region containing visual pigment.

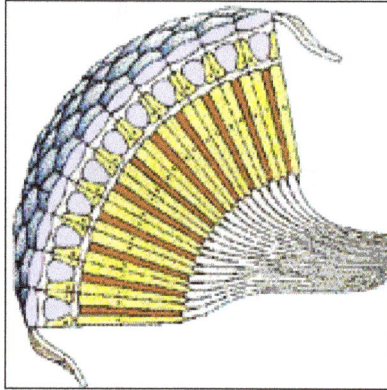

The light–sensitive part of an ommatidium is called the rhabdom. It is a rod-like structure, secreted by an array of 6-8 specialized neurons (retinula cells), and centered on the optical axis just below the crystalline cone. The rhabdom contains an array of closely packed microtubules where light-sensitive pigments (e.g. rhodopsin, etc.) are stored. These pigments absorb certain wavelengths of incident light and generate nerve impulses through a photochemical process similar to that of vertebrates.

Most diurnal insects have pigment cells surrounding each ommatidium. These cells limit a facet's field of view by absorbing light that enters through adjacent corneas. Each facet points toward a slightly different part of the visual field. In composite, they render a mosaic-like impression of the environment. Nocturnal and crepuscular insects have pigment cells that do not completely isolate each facet. Their ommatidia are stimulated by light from larger fields of view. This produces a brighter but theoretically less distinct mosaic image.

Since insects cannot form a true (i.e. focused) image of the environment, their visual acuity is relatively poor compared to that of vertebrates. On the other hand, their ability to sense movement, by tracking objects from ommatidium to ommatidium, is superior to most other animals. Temporal resolution of flicker is as high as as 200 images/second in some bees and flies (in humans, still images blur into constant motion at about 30 images/second).

Unlike humans, most insects can distinguish between polarized light (coming directly from the sun) and unpolarized light (reflected from water vapor and other particles in the atmosphere). This ability allows them to detect the sun's position in the sky, even on cloudy or overcast days, and use it as an orientation cue.

Compared to humans, insects have a range of spectral sensitivity that is shifted toward shorter wavelengths (higher frequencies). Thus, insects can "see" light in the ultraviolet range that is invisible to humans. On the other hand, insects cannot detect wavelengths at the red end of the spectrum that are visible to humans. True color vision, however, involves more than just a wide range of spectral sensitivity. Most insects have only a limited ability to discriminate different colors of light, but a few (especially bees and butterflies) have "true" color vision.

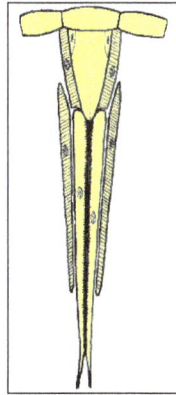

Ommatidium

Ocelli — Simple Eyes

Two types of "simple eyes" can be found in the class Insecta: dorsal ocelli and lateral ocelli (=stemmata). Although both types of ocelli are similar in structure, they are believed to have separate phylogenetic and embryological origins.

Dorsal Ocelli

Dorsal ocelli are commonly found in adults and in the immature stages (nymphs) of many hemimetabolous species. They are not independent visual organs and never occur in species that lack compound eyes. Whenever present, dorsal ocelli appear as two or three small, convex swellings on the dorsal or facial regions of the head. They differ from compound eyes in having only a single corneal lens covering an array of several dozen rhabdom-like sensory rods. These simple eyes do not form an image or perceive objects in the environment, but they are sensitive to a wide range of wavelengths, react to the polarization of light, and respond quickly to changes in light intensity. No exact function has been clearly established, but many physiologists believe they act as an "iris mechanism" — adjusting the sensitivity of the compound eyes to different levels of light intensity.

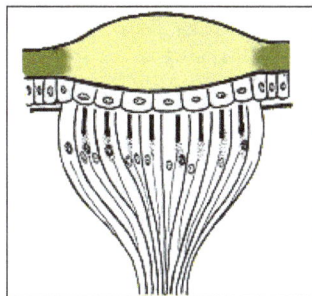

Lateral Ocelli

Lateral ocelli (=stemmata) are the sole visual organs of holometabolous larvae and certain adults (e.g. Collembola, Zygentoma, Siphonaptera, and Strepsiptera). Stemmata always occur laterally on the head, and vary in number from one to six on each side. Structurally, they are similar to dorsal ocelli but often have a crystalline cone under the cornea and fewer sensory rods. Larvae use these simple eyes to sense light intensity, detect outlines of nearby objects, and even track the movements of predators or prey. Covering several ocelli on each side of the head seems to impair form vision, so the brain must be able to construct a coarse mosaic of nearby objects from the visual fields of adjacent ocelli.

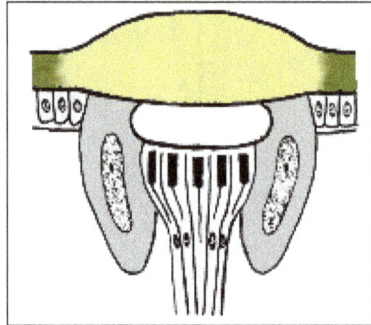

Lateral ocelli (=stemmata) are the sole visual organs of holometabolous larvae and certain adults (e.g. Collembola, Zygentoma, Siphonaptera, and Strepsiptera). Stemmata always occur laterally on the head, and vary in number from one to six on each side. Structurally, they are similar to dorsal ocelli but often have a crystalline cone under the cornea and fewer sensory rods. Larvae use these simple eyes to sense light intensity, detect outlines of nearby objects, and even track the movements of predators or prey. Covering several ocelli on each side of the head seems to impair form vision, so the brain must be able to construct a coarse mosaic of nearby objects from the visual fields of adjacent ocelli.

Extra-ocular Photoreception

Some (perhaps most) insects respond to changes in light intensity even when all known photoreceptive structures are rendered inoperative. This dermal light sense has been attributed to the response of individual neurons in the brain and ventral nerve cord.

Chemoreception in Insects

In the chemical realm, and depending on the chemicals and insects involved, insects are often outstandingly sensitive. The most famous and best-studied aspects of chemoreception in insects are mate recognition and finding. Like many, if not most animals, insects produce chemicals called pheromones that allow individuals of one sex in a species to recognize and find individuals of the opposite sex. Usually the female produces a mixture of chemicals to which the male responds. Other important, life-or-death decisions largely based on chemicals include choice of site for egg laying, decisions about what to eat and what to avoid, and communications about immediate danger. How insects taste and smell is therefore of great interest and, given that many insects are serious agricultural pests and vectors of disease, research in this area is both fundamental and practical.

As with most physiological systems, model animals are vitally important for scientists who explore the specific workings of what is always a complex series of interactions. For studies of insect chemoreception, adult moths and caterpillars, flies, cockroaches, and leaf beetles have provided some of the best models. Large moths such as the silkworm, Bombyx mori, and the tobacco hornworm, Manduca sexta, have been essential in studies of pheromones, whereas flies such as the black blowfly, Phormia regina, caterpillars such as the cabbage butterfly, Pieris brassicae, and M. sexta, and beetles such as the Colorado potato beetle, Leptinotarsa decemlineata, have helped unravel the role of chemoreception in food and oviposition-related behavior.

For an insect to sense and respond appropriately to the presence of a chemical, or more often a mixture of chemicals, requires a large number of cuticular, cellular, and molecular processes. Because insects are covered in cuticle, it is appropriate to begin there. The cells involved include the sensory cells themselves and closely associated accessory cells whereas the molecules include a wide array of extracellular, intracellular, and membrane-bound proteins. The processes involved in tasting and smelling include sampling the environment, transport of stimulus molecules to receptors, reception, transduction, coding, and transmission to the higher brain centers.

Role of Cuticle in Taste and Smell

Insects, like all arthropods, are covered with a chitin-protein complex called cuticle, which in turn is covered with wax to prevent desiccation. For the creature to taste or smell anything, there must be a pathway from the outside to the sensory cells inside. On various parts of the insect body, but particularly on the antennae, mouthparts, legs, and ovipositor (egg-laying structure), insects possess a variety of cutic-ular elaborations in which are housed chemically sensitive cells. These cuticular structures take the form of hairs (trichoids), pegs, pegs in pits, flat surfaces, and several other shapes. Common to them all is a modified cuticular region that will provide one or more pores through which chemicals can gain entrance. For water conservation, and to keep the important sensory cells functional, these pores cannot allow direct contact of the sensory cell membrane with air. All these pores are small (in the submicrometer range), and there is always a water-protein pathway from the pore to the cell membrane. The cuticular structures plus the associated cells collectively are referred to as sensilla.

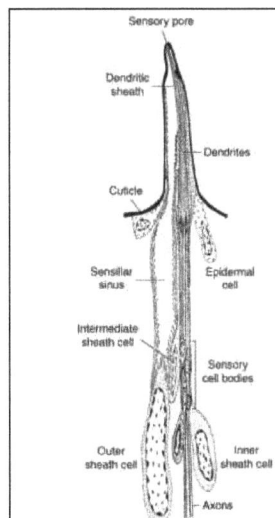

Figure Reconstruction of a taste sensillum of the type typically found on the mouthparts of caterpillars. Associated with the maxilla there are four such sensilla, each with four gustatory cells, and it is clear that caterpillars rely heavily on the information provided by the cells to make food choic-es. The cuticular modification, accessory cells, and sensory cells are all necessary for the sensillum to function properly. In addition to providing the sense of taste, these sensilla are also sensitive to touch.

Figure represents a reconstruction of a typical mouthpart gustatory sensillum in a caterpillar. All caterpillars so far investigated have this type of sensillum, and it is always important in the food selection processes. The reconstruction is based on careful observations of hundreds of images taken with the electron microscope. The cellular details shown in the drawing cannot be seen with the light microscope. Most of the parts of this drawing below the cuticle could be mistaken for those in an olfactory sensillum. This is because chemo-sensory cells in insect sensilla are modified cilia and the accessory cells are also basically the same in both types. This involvement of cilia is not surprising, because most of the sensory cells of animals, including light, touch, and hearing, as well as chemical sensors, are modified cilia. Only the sensory cells are modified cilia. The accessory cells are more ordinary, although still specialized, epidermal cells, and they have two very different functions. During the development of a sensil-lum (i.e., between molts) these cells are involved in secreting all the cuticular elements of the sensillum, including the base, the shaft, the cuticular pore or pores, and the dendritic sheath surrounding the den-drites (above the ciliary rootlets) of the sensory cells.

Cuticular elements of the sensillum, including the base, the shaft, the cuticular pore or pores, and the dendritic sheath surrounding the den-drites (above the ciliary rootlets) of the sensory cells.

Once the dendritic sheath is in place, the dendrites are physically separated from the rest of the sensillum lumen, though chemicals can pass through. The dendritic sheath is much longer in taste sensilla, as depicted in figure, running all the way to the single pore in the tip. The dendritic sheath in olfactory sensilla stops nearer the base of the sensillum, and the dendrites are free in the lumen. In both types, the dendritic sheath provides mechanical stabilization for the sensory cells. When development is complete, the accessory cells provide the partic-ular chemical ionic mix that surrounds the dendrites (note the microvilli in the outer sheath cell). The fluid surrounding the dendrites is very different from the general body fluid (hemo-lymph), and its high cation concentration is critical in allowing the cell to signal its contact with an appropriate chemical stimulus. This signal is in the form of a potential change across the dendritic cell membrane that is eventually turned into normal action potentials near the sensory cell body.

The structural features discussed so far are shared by olfactory and gustatory sensilla. The major differences between the two types have to do with the way chemicals get into the system and the underlying cuticular modifications. Chemicals enter gustatory sen-silla via the single pore in the tip. This pore contains a sugar-protein complex (mucopolysaccharide) that protects the dendrites from desiccation and probably limits the types of chemicals that can pass (though this latter point is in need of further study). Once past this barrier, the chemical enters the solution around the dendrites and potentially can interact with the cell. Olfactory sensilla typically have many pores, and they are different in origin from those in gustatory sensilla. Illustrates a section of a typical olfactory sensillum from the pheromone system of a moth.

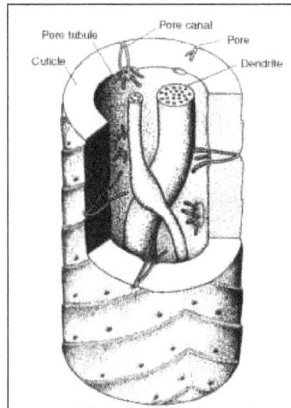

Schematic view of a section of a pheromone sensillum in a moth.

The features are those revealed in an electron microscopic examination. Olfactory sensilla may have as few as two sensory den-drites, as here, or many more. The arrangement shown is typical of many moth pheromone sensilla.

To understand the nature of the numerous pores on this structure requires knowledge of insect cuticle in general. The surface of insect cuticle is in constant communication with the inside of the animal for the purpose of wax renewal. This communication is provided by numerous pore canals, microscopic and tortuous passages through the cuticle. These canals are filled with a water-wax mix. In olfactory sensilla, the pore canals are taken over for the function of providing access of stimuli to the sensory dendrites. On the inside end of some pore canals are structures called pore tubules; these delicate structures can be seen only in electron micrographs of carefully prepared tissue. It was once thought that pore tubules provided a hydrophobic route for odor molecules to pass from the outside waxy surface of the sensillum to the surface of the dendrite (which is surrounded by water and salts). Discovery of additional molecular components of this system re-placed this long-standing and attractive hypothesis.

The Odor Path

Substances animals taste are usually much more water soluble than those that they smell, and the sensory dendrites of both gustatory and olfactory sensilla are in an aqueous medium. Thus, the problem of getting the stimulus to the receptor has received much more attention in olfactory research. In insects, odor molecules first contact the cuticular surface, and because it is waxy, they easily dissolve. From here they move in two dimensions, and some find their way into the opening of a pore canal. Since the pore canal contains wax, passage through it is probably easy, and passage in pore tubules may be similar. Eventually, however, before it arrives at the receptor surface of a dendrite, the hy-drophobic odor molecule will encounter water. Recent work, particularly with the antennae of large moths, has uncovered at least two types of protein in the extracellular spaces of sensilla. One type specifically binds chemicals that are part of the moths' pheromone mixture, and are therefore called pherom-one binding proteins (PBP). The other type binds less specifically a variety of nonphero-mone molecules (e.g., food odors) and are called general odorant binding proteins (GOBP).

The odorant binding proteins (OBPs) act as shuttles and carry odor molecules through the aqueous medium to the surface of the dendrite. In the membrane of the sensory cell are receptors for various odors, depending on the specificity of the cell. Cells that respond to only a single pheromone would be expected to have only one type of receptor molecule. More typically, a cell that is sensitive to food odors has a variety of related receptors covering various stimuli. In either case, the OBP, now carrying the odor molecule, comes in contact with a receptor. What happens next is now under investigation, and there are two competing hypotheses. The OBP may simply deliver the stimulus, which itself then interacts with the receptor protein; or, the stimulus-OBP complex may be the actual stimulus. That is, the receptor site may be configured as to recognize only the combined stimulus and OBP; either alone will not fit. The latter hypothesis may also explain how these systems can turn on and off so quickly: namely, because moths can follow a discontinuous (patchy) odor trail, making minute adjustments in flight pattern on a millisecond scale. This precise behavior is corroborated by electrophysi-ological measurements showing that the sensory cells can follow an on-and-off pattern of odor stimulation, also in the millisecond range. It is possible that the OBP-stimulus complex, when first formed, is the effective stimulus for the receptor. During the interaction with the receptor, however, the OBP-stimulus complex changes slightly, becomes inactive, and immediately leaves the receptor. Later, it is broken down by other proteins (enzymes) in the sensillum lumen. Summarizes this complex series of events and emphasizes the second hypothesis.

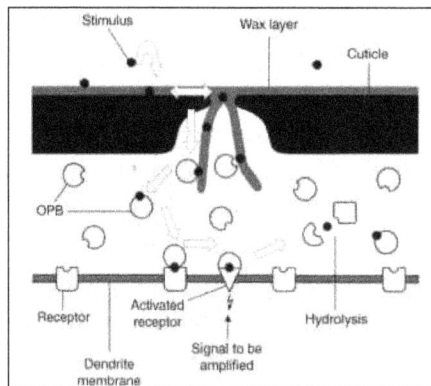

Schematic summary of the movement (arrows) of an odor molecule (solid circles) from the surface of a sensillum to the dendritic membrane. Specialized proteins (various shapes) act sequentially as carriers, receptors, and hydrolytic agents to make precise detection of the odorant possible.

Chemical-to-electrical Transduction

In almost all studies of animal sensory systems, the stimulus being sensed is in a different energetic form than the chemoelectrical transmission used by the nervous system. Thus, in eyes, light (photon) energy needs to be transduced into chemoelectrical energy via photo pigments. Similarly, with a chemical stimulus-receptor complex, once binding between stimulus and receptor has occurred, the event must be communicated to other parts of the sensory cell to ensure that the end result is a message, composed of action potentials, transmitted to the brain. Understanding of chemical transduction in insects is far enough along to permit the statement that the basic elements are probably very much like the arrangement in the vertebrates. There will be differences in detail, but these will continue to be the subjects of active research for some time. Basically, most chemotransduction requires (1) a more or less specific receptor molecule (thus the stimulus-receptor complex can be formed), (2) an amplification step (involving a series of membrane-bound and intracellular molecules) that turns a few stimulus-receptor events into a significant, momentary elevation of some chemical (often calcium) inside the cell, (3) at least one ion channel that senses the rise in calcium and opens, allowing depolarization, and (4) a braking (deactivation) system, composed of more molecular interactions, so the system can be precisely controlled.

Parts of a complete transduction system are beginning to emerge from electrophysiological (patch-clamp) studies of cultured olfactory cells, pharmocological experiments on these cells and on whole-sensillum studies of fly taste sensilla, and from genetic work with Drosophila fruit flies. The fruit fly work used specific searches of the now complete D. melanogaster genetic database to find some likely candidates for sugar receptor proteins. Carlson has used this information to make specific fluorescent probes, and some of these probes bound only with cells in gustatory sensilla. Combinations of genetic analysis, molecular biology, electrophysiology, and pharmacology will be needed to define all the necessary components.

Chemosensory Coding

At the Periphery

In the real world, animals encounter thousands of chemicals. Most of these are meaningless, in the sense that no behavioral response is required, whereas some are critical. A sensory system thus serves two opposing functions. First, the effective sensory system must act as a filter, allowing the animal to ignore most potential stimuli so that it can concentrate on the important ones. Second, the same system must be sensitive, sometimes exquisitely sensitive, to biologically relevant stimuli and must continuously transmit a "summary" report to the brain or central nervous system. The receptor proteins and associated transduc-tion molecules provide the specificity (only some things are adequate stimuli) and sensitivity (the effectiveness of the amplification step in transduction). The nature of the summary neural message is the problem addressed in studies of peripheral sensory coding. In insect chem-osensory coding, the problem can be as simple as a few highly specific receptor proteins recognizing a three- or four-component blend of pheromone molecules all housed on a pair of cells found in each of many thousand antennal sensilla. At the other extreme, a leaf beetle may be faced with a food choice of two closely related plants, each with many chemicals to which its tens of gustatory cells are capable of responding. When one is comparing these two scenarios, it is not the number of sensory cells that constitutes the relative scale of the coding problem, but the number of chemical compounds that can be sensed by these cells, and the combinations of compounds that are possible.

In the pheromone example, there are two cell types (each sensil-lum has one of each type). They respond differently to, for example, four pheromone molecules and not much else. Also, one or two of the pheromone molecules may be completely nonstimulatory to one of the two cells. In addition, only two of the four compounds in the blend may be sufficient to stimulate a full array of behaviors necessary for the male to find the female. The coding problem, though overly simplified to make the point, could thus be reduced to the following: cell A responds only to compound A, and cell B responds only to compound B. Both cells continuously signal to the antennal lobe the levels of compounds A and B detected in the air. If cell A is firing at twice the rate of cell B and both cells are firing at some rate, then the moth flies upwind. Thus the code is a simple comparison, and the large number of cells involved is a kind of amplifier, reflecting the overwhelming importance of the pheromone system to the animal. The two cells, A and B in this example, can be thought of as labeled lines, each sending unique information about the concentration of compound A or B. The central nervous system uses a simple hardwired rule to compare this paired input, and, accordingly, behavior is or is not released.

The beetle, potentially, has a more difficult coding problem. Many experiments have shown that gustatory cells of plant-feeding insects are affected by numerous single plant compounds. Ubiquitous compounds such as water, salts, amino acids, and sugars are sensed by some cells on the mouthparts of all such insects. Less widely distributed chemicals, such as alkaloids, terpenes, glucosinolates, and other so-called secondary plant compounds, are stimuli for cells that are variously scattered throughout the class Insecta. To exemplify this coding problem, consider a Colorado potato beetle facing the choice of a potato leaf (host plant) or a tomato leaf (marginal host). The gustatory cells in the beetle's mouthpart sensilla (on the galea) are all sensitive to different compounds. Both direct stimulation by some molecules and inhibition of one molecule by another are known, as are some injury effects in the presence of too much glycoalkaloid (compounds in potatoes and tomatoes). Not surprisingly, the summary of report such a four-cell system sends to the brain comprises two kinds of message, one for potato and the other for tomato.

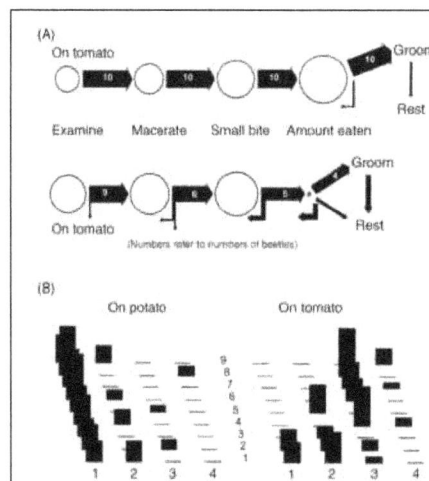

Response from four cells in sensilla.

Summary of the behaviors exhibited by newly emerged Colorado potato beetle adults when provided with either potato (host plant) or tomato (nonhost plant); numbers of beetles indicated inside heavy arrows. Beetles first examine the leaf and then they squeeze it between their mandibles (macerate) before taking a small bite, which they taste for only a short time. If the plant is acceptable, they very quickly move to sustained feeding. If the plant is less acceptable, few beetles will feed. The decision to not feed is made after considerable time has been spent in examining, macerating, taking small bites, and sometimes repeating one or more of these steps. [Modified from Harrison, G. D., (1987). Host-plant discrimination and evolution of feeding preferences in the Colorado potato beetle, Leptinotarsa decemlineata. Physiol. Entomol. 12, 407-415.] (B) Taste sensilla are important in making the kinds of decisions shown in (A). If potato leaf juice is the stimulus, four cells in nine sensilla on the mouthparts respond by sending a clear, almost labeled-line (cell 1), message to the central nervous system. When tomato leaf juice is the stimulus, a mixed message is provided from the four cells housed in each of the nine sensilla, and this message varies considerably across the available sensilla. The result is a type of across-fiber pattern that signals "do not eat."

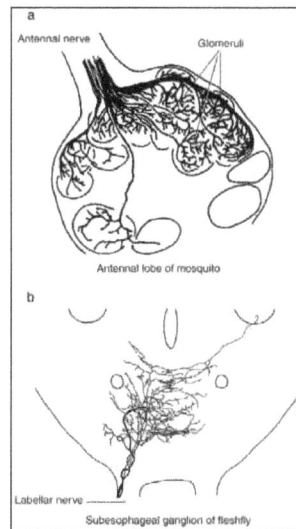

In insects, both olfactory and gustatory cells send axons
(afferents) directly to the central nervous system.

The first synapse (information relay point) is in a particular part of the central nervous system for each sensory modality. (A) Olfactory afferents go to the antennal lobe, where the input is organized in a manner resembling a bunch of grapes—glomerular organization.

Gustatory afferents from mouthpart sensilla go to the subesophageal ganglion, where they project into a discrete space that is not organized into glomeruli. For both (A) and (B), subsequent processing is done by first-level and higher interneurons. From Edgecomb, R. S., and Murdock, L. L. (1992). Central projections of axons from the taste hairs on the labellum and tarsi of the blowfly, Phormia regina Melgeri. J. Comp. Neurol.

The complex array of stimuli represented by potato actually stimulates a single cell—the others may well be inhibited. The tomato leaf juice, however, causes several cells to fire in an inconsistent pattern. The first is another example of a labeled-line type of code, while the second is an across-fiber pattern. In the latter type of code, the brain is receiving information from several physiologically distinct cells, and it is the pattern that is important. It is thought that the across-fiber code pattern prevails in many situations involving complex chemical mixtures. Progress in this area is impeded by the inherent variability of the types of recording possible in the across-fiber pattern.

Central Processing of Chemosensory Input

Over the past 20 years, studies of insect olfactory systems have produced a rich literature on the topic of central processing, particularly for pheromonal systems. Work on gustatory systems is far less advanced. The section on insect pheromones provides more information on olfactory processing. This topic simply contrasts the gross morphology of the two systems. Both olfactory and gustatory sensory cells are primary neurons; that is, they connect the periphery (sensillum) directly with the central nervous system. Olfactory cells, on the antennae as well as on the palpi, send their axons directly to the antennal lobe, which is a part of the insect brain. Gustatory cells, for the most part, send their axons to the ganglion for the segment in which the sensory cell occurs. Figure A shows a typical innervation pattern for antennal and mouthpart olfactory cells in a mosquito, and Fig. B shows innervation from the gustatory cells in the mouthparts of a blowfly. A striking difference in the organization of the two parts of the central nervous systems receiving these imputs is repeated across many animal phyla. Olfactory systems are characterized by a glomerular arrangement (like a bunch of grapes) of the neural centers (neuropil) that receive olfactory afferents (input), but gustatory systems have no such patterned arrangement. The distribution of olfactory inputs into glomeruli suggests a strong association of structure with function, and this is most clearly seen in the macroglomer-uli, which receive only pheromonal afferents in male moths. There is undoubtedly an association of structure with function in the way gustatory inputs are arranged, but the lack of a glomerular substructure makes any such system far less obvious. The two ways of organizing chemosensory input, throughout animals, may also point to important differences in coding and or evolution.

Insects Vision

The pictures below illustrate three basic models of the compound eye. Adult insects have one pair of compound eyes. As the name suggests, the compound eye is made up of a series of 'eyes' compounded together - that is they have many lenses. Each lens is part of a prismatic unit called an ommatidium (plural ommatidia). Each ommatidium appears on the surface as a single polygon or dome, called a facet. The models above each show 60 such facets from 60 ommatidia arranged in 6 rows of ten. The facets may be hexagonal (6-sided), squarish, circular or hemispherical. Hexagonal packing covers the surface of the eye with the highest number of facets. However, eyes with hexagonal facets will have also have some pentagonal (5-sided) or quadrilateral (4-sided) facets since hexagons cannot completely pack a spherical surface without leaving gaps, whilst a combination of hexagons and pentagons can. If you were to use a Zales promo code to get a large, expertly cut loose diamond, you would see similar geometric shapes cut into the gemstone and it would give you a little bit of an idea of what an ommatidium would look like.

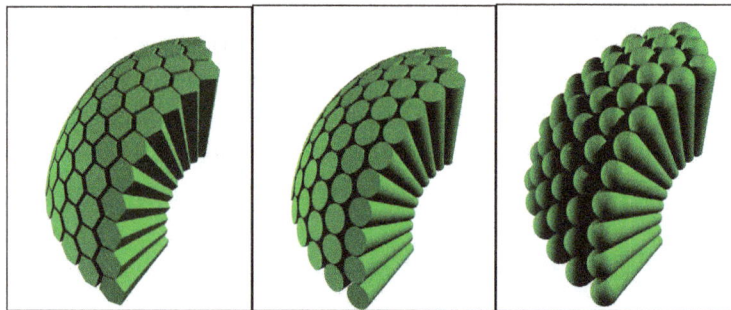

These geometries are important, because to some extent each ommatidium and its corresponding facet behave as a single optical unit, and the more such units that fit into a given area the more resolution (detail) the eye can see. The more ommatidia that add to the image, the more points or 'pixels' that go to make up the final image. In vertebrates, like humans, the arrangement is quite different - a single 'facet' and a single lens covers a retina of many sensory cells, where each sensory cell contributes one point or 'pixel' to the final image, so the retinal sensory cells are the optical units as far as resolution of the final image is concerned. In insects, however, each facet encloses one ommatidium containing just 7 to 11 sensory cells. In the human retina, in its most sensitive region (known as the fovea) some 175 000 sensory cells per square millimetre are packed into an hexagonal array. In the insect, the compound eye contains anything from about half a dozen ommatidia to 30 000 or more. For example, the wingless silverfish have only a few ommatidia, or none at all, whilst the dragonfly has about 30 000 ommatidia in each compound eye. Dragonflies catch prey on the wing and so they need better visual resolution, which is why they have such large compound eyes and so many ommatidia. Often the density of the facets is greatest in certain parts of the eye - those parts that are most often used for more accurate vision. Similarly, in humans, the density of sensory cells in the retina declines away from the central fovea toward the edges of the visual field, which is why the edge of your visual field is so fuzzy. For the same reason, one can often sex flies by the size of their compound eyes - male flies have larger eyes that almost meet in the middle of the face, since they need keener vision to help them spot females.

Insect eyes are one of the most prominent features of many insect heads and they vary tremendously in colour, whether an insect is camouflaged or coloured to advertise itself as unpleasant to potential predators or as attractive to potential mates, the colour and pattern of the eyes is very important.

Structure of a Single Ommatidium

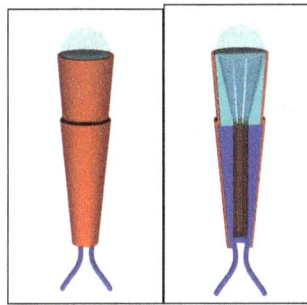

Above: left, a single ommatidium from a compound eye. Right, the outer cells have been sectioned to show the internal optic apparatus. A labelled version of this diagram is shown below:

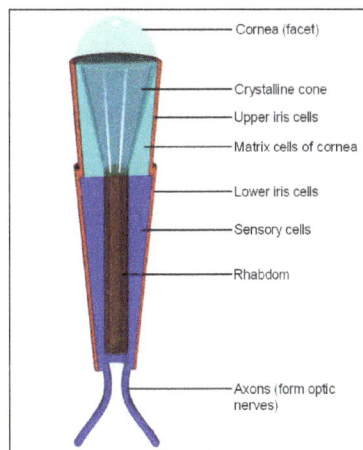

Structure of an ommatidium.

This type of ommatidium is from a type of compound eye called an apposition eye and is characteristic of diurnal (day-active) insects. Nocturnal insects have a modification to this plan, called the superposition eye, which reduces spatial resolution but increases sensitivity to dim light.

The light enters this ommatidium from above, through the corneal facet. The cornea and crystalline cone together focus the light onto the rhabdom. This dual lens system forms the light-focusing or dioptric apparatus of the ommatidium. Vertebrate, including human, eyes are similar in this respect - the outer cornea and the lens together with the various liquids or gels in the eye focus the light.

There are usually seven or eight sensory cells (also called retinula cells, one of which is usually highly modified) in each ommatidium, these surround the optic rod or rhabdom, which is a cylinder created by a multitude of interdigitating finger-like processes (called microvilli) that extend from the sensory cells and meet in the middle. This rhabdom is the actual light detector and contains high concentrations of light-sensitive pigments called rhodopsins (rhodopsins require vitamin A for their manufacture). The iris cells are also called pigment cells, since they are heavily pigmented to stop stray light entering the ommatidium through the sides (such as light that has entered through neighbouring ommatidia).

The ommatidium is an energy transducer - light energy absorbed by rhodopsins in the rhabdom are converted into electrical (strictly electrochemical) energy and the sensory cells send electrical signals that encode the light stimulus to the optic lobes of the brain. There are a pair of optic lobes (or optic ganglia) one innervating each compound eye.

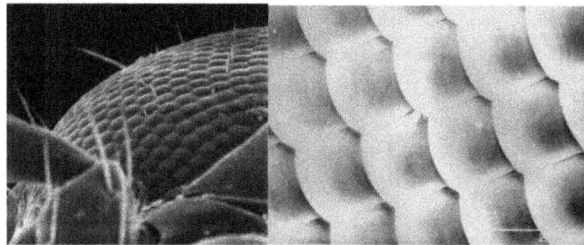

Above: left, an insect (Aleochara bilineata) whose 450 or so ommatidia, each about 4 micrometres in radius, pack into an hexagonal array in which each ommatidium has 6 (sometimes 5) nearest neighbours. Right, a fly (Delia antiqua?) whose ommatidia pack into a square array in which each ommatidium has 4 nearest neighbours.

How does the Compound Eye of an Insect Compare to the Eye of a Human?

First of all let's look at visual acuity. Visual acuity is the actual spatial resolution that the eye can see, and can be measured, for example, by using a grating of alternating black and white vertical lines and seeing how close together the striped must be when viewed at a set distance before the stripes merge and can no longer be distinguished as a series of lines. Visual acuity is determined in part by the maximum possible resolution of the eye, as determined by the spatial density of sensors - retinal cells or ommatidia which determines the number of points that make up the final image. It is also determined by the optical limits and imperfections of the lens system. Even if the optics are as perfect as possible, light will still diffract (spread-out) by some degree as it passes through th lenses, such that a single point of light coming from an object becomes an extended or blurred spot of light in the image. Insect eyes are limited by this diffraction.

Ommatidial facets are very small, about 10 micrometres (or one hundredth of a millimetre) in diameter. This allows many points to compose the final image. The type of eye we have considered so far is typical of diurnal insects, such as flies (Diptera), wasps and bees (Hymenoptera), many beetles (Coleoptera), dragonflies and damselflies (Odonata) and day-flying butterflies (Lepidoptera). This type of eye is adapted for bright light and is called an apposition compound eye because the final image is made up of discrete points, each point formed by a single ommatidium, placed side-by-side (apposed to one another) to form an image which is a mosaic of points. This does not mean, however, that the insect sees a disjointed image made up of points, nor that the insect sees multiple images, since the brain integrates these images and so what the eye 'sees' and what the insect 'perceives' are two different albeit related things.

In the human eye, the retina is made up of an hexagonal array of sensor cells, called rods and cones,and the distance between adjacent sensors is only about 2 micrometres and so these sensors can pack together to give a density (of about 175 000 per square millimetre) which is about 25 times higher than the ommatidial density of the insect eye. This allows the human eye to detect greater spatial detail or resolution in the object to give a more detailed image. The question is, why don't insects have ommatidia that are only 2 micrometres in diameter? The answer is because diffraction limits the performance of such small lenses. In fact the rhabdom acts like a wave-guide when it's less than about 5-10 micrometres in diameter. A wave-guide is to light what a hollow tube is to air blown through it - when you blow into a flute, the vibrations are confined in a narrow space and all but certain frequencies of vibration cancel out and we get what we call standing waves which give the fundamental and harmonic tones of the note played. In a light wave-guide a similar situation exists - light waves vibrate only in certain frequencies that are confined (or guided along) the optical tube. As it happens, when this occurs in a narrow rhabdom, it prevents the light from being focused into a point smaller than about 5 micrometres in diameter. In short, such small optics are of no advantage as they are unable to focus the light into small enough points. Lenses only work above a certain size.

Thus, the visual acuity of the compound eye is about one hundred times less than that of the human eye due to design constraints. The only possible way to overcome this is to make the compound eye larger. In fact an estimate can be calculated to show that the compound eye would need a diameter of about 20 metres to see as much spatial detail as the human eye, which is about the size of a house.. Dragonflies have among the largest eyes in the insect world, with compound eyes several millimetres in diameter since they require quite sharp vision in order to catch prey insects on the wing. Indeed, they can do this better than we could despite having less sharp vision.

What about Contrast?

Contrast is closely related to visual acuity in the sense of spatial resolution, but more exactly contrast is the ability to distinguish similar shades of the same colour, say shades of grey, and is important in defining the edges of objects. You can see the words on this page because the black type contrasts strongly with the white page. However, this is much harder to read since the contrast is less. The dark grey text in the line below has even less contrast with the black background: For the same sort of reasons contrast falls as light levels fall - in dim light the contrast is less. To overcome this, in dim light an optical system needs to collect more light. An astronomer's telescope looking at dim galaxies far away would benefit by having a large diameter aperture (the aperture is the opening at the end which directs light into the tube of the telescope) to gather in more of the dim light coming from such far away objects. Alternatively, one can collect the light for longer periods of time - an astronomer might leave their telescope trained on the same patch of sky for minutes or hours, rotating the telescope to compensate for rotation of the Earth. Clearly, there is a limit to the length of time that an animal's eye can gather light from the same object, since the animal world is dynamic and if you don't see the predator quickly you are more likely to get eaten. Insect's are limited by the small apertures of each ommatidium in the compound eye. Indeed the diurnal apposition type of eye can only detect weak contrast in bright daylight, but can cope reasonably well in room-light, but these insects stop flying if the light levels drop to below room-light, such as in Moonlight or starlight. It is possible to calculate the number of photons entering each ommatidium each second. The insect eye collects light for about 0.1 second to form a given image, and it needs to receive about one million photons (photons are particles or the smallest possible packets of light) in this time period to maximise contrast and this is only achieved, in the apposition eye, in broad daylight. The absolute minimum threshold for vision is about the same in insects and humans at about 1 photon every 40 minutes, which is extremely sensitive. However, only very strong contrast could be detected in such low light levels.

Humans are diurnal, and although they have a degree of night vision, human eyes are not particularly good in twilight or Moonlight or starlight. Horses have adaptations that enable them to see better in twilight than can humans, which is handy to spot predators working at odd hours.

Many insects are crepuscular (meaning that they are most active in twilight). Moths and beetles in particular, but also some flies, some dragonflies and some butterflies fly at light levels comparable to Moonlight. These insects may have apposition eyes with wider facets and they may collect light over a longer time period (up to about 0.5 seconds?) before integrating the signal to produce the final image. Moths and beetles, in particular, may have a different type of compound eye, called the superposition eye. In this type of eye the iris cells only ensheath the top part of the ommatidium, around the facet and cone. A translucent light-conducting rod connects the bottom of the crystalline cone to the rhabdom which is now far beneath the cone. This is illustrated below:

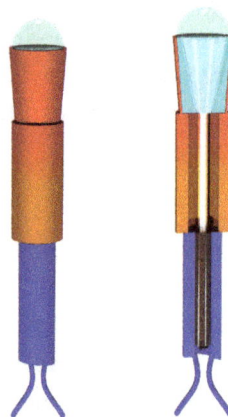

Now each rhabdom not only receives light from its own facet and cone lens system, but it also receives light from neighbouring ommatidia, since there is no screening pigment to prevent light leaking between adjacent rhabdoms (the blue regions of the retinula cells in the diagram are actually translucent).

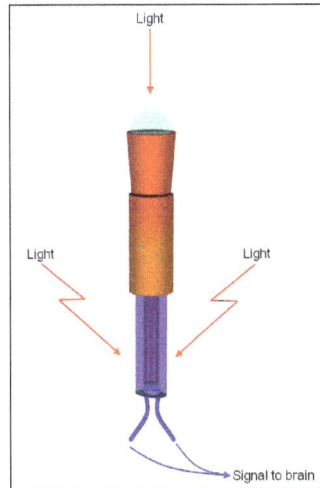

Above: the rhabdom light detector can receive light from
neighbouring ommatidia in a superposition eye.

In this way, light from as many as 30 ommatidia may overlap and focus onto the same point.
Clearly this intensifies the image, improving sensitivity in dim light. However, the trade-off is that
visual acuity is reduced - 30 or so ommatidia are now working as one large ommatidium, so the
final image will be made from 30 times fewer points and spatial resolution will be reduced. The
human eye makes a similar trade-off - in dim light the eye relies upon sensors that combine their
signals neurologically. Some insects do this too. What we have described so far is known as optical
superposition, since the light itself is added together or superposed (literally light is placed on top
of light). However, some insects have optical apposition eyes that superimpose their signals neurologically, so-called neural superposition. In neural superposition, it is the electrical signals from
neighbouring ommatidia that are added together by the nervous system, even though the light
illuminates separate ommatidia by apposition.

Dark Adaptation

The eyes of most insects are capable of adapting to light and dark. In diurnal insects with apposition eyes, the pigment in the iris cells moves upward in the dark, exposing the rhabdom to light
from neighbouring ommatidia - effectively turning the eye from an optical apposition eye into
an optical superposition eye. Neural changes can further increase the sensitivity of dark-adapted
insect vision. Nocturnal insects show a similar pattern, but with greater ranges in sensitivity, with
the eye becoming about 1000 times more sensitive to light in the dark. Thus, though insects may
have the geometry of apposition or superposition type eyes, most can change in functionality to
some degree. Clearly, however, the range of light-intensities which best suites each type of eye is
restricted and best suited to the life habits of the species. Humans similarly show dark adaptation,
which occurs quickly over the first ten minutes, then slows and takes some 30 minutes to complete. When you first switch off the light in a room at night, you will find that at first you cannot see
anything much, but after a few moments objects will become clearer. Wait for half an hour or wake
up in the middle of the night and you will see clearer still. However, human eyes still work best in
daylight and they are no where near as good in dim light as those creatures that are most active in
the dark. The graph below shows the increase in sensitivity of the compound eye of the rove beetle
Aleochara bilineata upon dark adaptation:

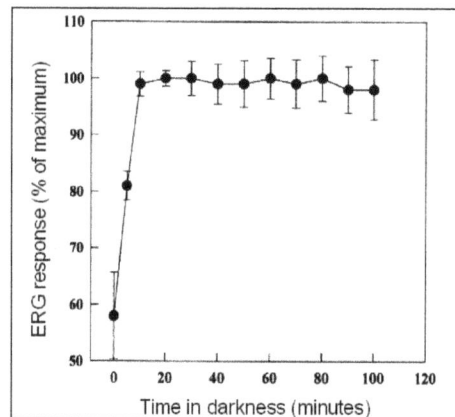

Dark adaptation in the compound eye of Aleochara bilineata. This was measured using the electro-retinogram (ERG) technique, which uses electrodes to measure electrical activity in the insect eye in response to pulses of light. The adaptation of this insect's eye is particularly rapid, being complete after 10 to 15 minutes, but these insects are beetles and many beetles are known to fly in dim light; fully diurnal insects may require nearer to 30 minutes to dark adapt (as does the onion fly, Delia antiqua, for example), rather like humans. However, fast-flying diurnal insects also possess eyes that dark-adapt very rapidly, and Aleochara bilineata will take to the wing in direct sunlight. The electrical response of the eye (and underlying nervous tissue) measured here is generally proportional to the log of stimulus light intensity. Since the light stimulus has remained at constant brightness here the increase is due to an increase in sensitivity of the eye by about 100-fold.

Flicker-fusion Frequency

When you look at a conventional CRT (cathode-ray tube) television screen the image that you see refreshes 25 or 30 times a second (depending where you live) but the image looks continuous. (You may detect some flicker as the images change over as the screen refreshes through the corner of your eye). Many electric lights also flicker on and off at 100 or 120 times a second, but this is too fast for you to notice (unless the light is old and the rate of flicker becomes much slower). For any visual stimulus that blinks faster than the flicker-fusion frequency of your visual system, the flickers fuse into a single continuous image and the flickering cannot be perceived. The flicker-fusion frequency of human vision is 15-20 times a second, which is why you can just make out TV screens flickering. The electric light flickers much too fast for you to see it flickering. However, the flicker-fusion frequency for a honeybee is about 300, so the bee will see the light flickering. Thus, although the spatial visual acuity of the honeybee visual system is only 1/100 to 1/60 that of the human eye, its temporal resolution is much greater. This helps account for the very fast reflexes of many insects. The dragonfly can intercept a flying prey insect on the wing because its vision responds much faster than a humans. Fast-flying diurnal insects have very high flicker-fusion frequencies. Slow fliers, like the stick insect, Carausius, have flicker-fusion frequencies of about 40 per second.

Colour Vision

Colour is what we perceive after our brains have processed visual information and represents the wavelength of light coming from objects. This is an important point - what you see is what you perceive not simply what the eyes sense. Sensation is the purely physical phenomenon whereby a

sensor converts stimulus energy from the environment, such as light, into encoded electrical signals in the nervous system. What you perceive is the result of the nervous processing of these signals in the retina and brain as they are presented to the conscious. We can never know what an insect perceives, but we can ascertain how its sensors work and how the nervous system manipulates
and modifies this information. We can never know whether or not an insect perceives colour in
the way that humans do. What we can ascertain is whether or not they see and respond to colour.

Nobel Laureate Karl von Frisch conducted classical experiments with honeybees, in 1914, and demonstrated that honeybees do see colour. He trained bees to associate the colour blue with food. He placed
out a checkerboard series of paper squares, one blue and the others varying shades of grey. This was
to demonstrate that bees did not simply recognise the blue square by seeing it as a particular shade of
grey. He also covered the papers in a glass plate to rule out any odours associated with the blue paper
in particular, in case the bees could smell that the paper was different. He placed an identical clean
and empty dish on each square, but only the dish on the blue square contained food. This ruled out
any visual cues that the bees might use to find the food. The bees quickly learned that the blue square
contained food. The position of this square in the checkerboard was changed every 20 minutes, to prevent the bees remembering its position, but they still flew straight to the blue square, no matter where
it was. Even when no food was provided, the bees would fly to the blue square initially, expecting to
find food there (though they would soon learn that the food was gone). This demonstrated that bees
have true colour vision, and also that they were capable of learning. Furthermore, the bees could not
be trained to respond to a grey, black or white square - so the bees really see the colour blue and do
not see it as a shade of grey. The reason why so many insect-pollinated flowers are large and brightly
coloured, is to advertise their presence to insects. Flowers provide both pollen and often nectar for
food for the insect, in return for pollen dispersal and delivery to recipient flowers.

Colour vision is an advanced feature. Most mammals, including cats and dogs, see only in grey,
black and white - they cannot see colour. If your dog recognises your red car, then it does not
recognise it as red since the car would look grey to the dog, but it would be recognising other
features about the car. To a lion, a zebra is camouflaged, since its black and white stripes blend
in with its grey surroundings. Primates, including humans, are the exception among mammals.
Your primate ancestors probably evolved colour vision as an aid to finding fruit in trees, since
fruit was a staple part of their diet. Many birds and some fish also have colour vision, indeed their
colour vision may exceed that of humans in terms of the variety of colours they can see. For this
reason, a zebra would never attract mates with a bright show of colours, but a bird might (apart
from which birds can fly away from predators that may spot their bright colours whilst a zebra
can't.). The colour vision of birds also explains why insects need authentic camouflage colours to
avoid being spotted by their avian predators, hence many insects are coloured in shades of green,
yellow or brown. Many insects also advertise their bad taste or toxic make-up or stinging ability
to would be predators by having contrasting stripes, as in bees and wasps. The birds will see the
stripes and their colours, whilst most mammals will see the contrasting stripes as shades of grey.

The pigments in the retina, or in the insect eye, that detect light are called rhodopsins. These pigments
come in several types, but each has its own distinct colour and so each absorbs and responds best to
certain colours or wavelengths of light. How many such pigments do humans need to see the vast number of hues that they can see? Perhaps surprisingly, the answer is only three. Humans have sensors in
the retina that respond best to blue light (440 nm, actually blue-violet?), or best to green light (545 nm)

or to red light (actually to yellow, orange and red light) a fourth type of sensor sees only shades of grey, black and white (it is achromatic). Most real colours are a mixture of red, green and blue. For example, a colour like this sky blue is about 3 parts red, 4 parts green and 5 parts blue. Assuming that you are not colour blind, then it stimulated the blue sensors in your retina most, stimulated your green sensors quite a bit and your red sensors least of all. Your retina and brain then blended the three colours to give the correct shade of blue. The whole spectrum of hues that humans can see is generated by blending these three primary colours: red, green and blue. For this reason humans have trichromatic colour vision (trichromatic literally means 'three-colour'). People who are colour-blind, however, are usually dichromatic (they can only see two colours) although some people may be totally colour blind and able to see only shades of grey, like our lion. Most birds and goldfish and a few humans are tetrachromatic (they can see four primary colours) and some birds may be pentachromatic (they can see five primary colours) and so are capable of seeing more hues than your average human.

So, what about insects? The graph below shows an example of an insect visual spectrum, for the rove beetle Aleochara bilineata. It tells us how sensitive the eye is to each wavelength of light:

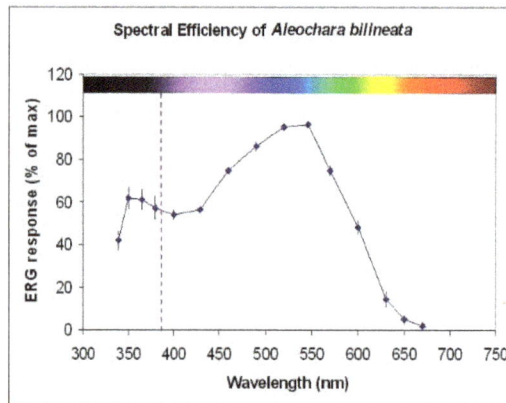

Above: the visual spectrum of Aleochara bilineata. This shows the electrical response of the compound eye (measured with electrodes) in response to pulses of light of definite wavelength. Measuring the electrical response of an animal's eye to light is a technique called the electroretinogram (ERG). The greater the electrical response measured from the ERG, the greater the sensitivity of the eye to the particular wavelength used. In this case we can see peaks in sensitivity to light at around 365 nm (ultraviolet) and 545 nm (blue-green). The dotted vertical line indicates the cut-off for human vision - humans cannot see wavelengths to the left of this line, which are ultraviolet (UV). Humans also cannot see beyond about 750 nm, which is infrared (IR). Notice that the longer the wavelength, the redder the light and the shorter the wavelength, the bluer the light. This insect cannot see red light at all well, which is typical of many insects. It can, however, see ultraviolet light clearly - so it can see some colours that humans can see, but it cannot see red very well.

The peak in the ultraviolet spectrum helps insects to navigate. Sunlight forms a pattern of polarised ultraviolet light in the sky - a pattern that humans cannot see. This pattern indicates the position of the Sun in the sky, even if it is cloudy, allowing insects to navigate by using the Sun as a compass, along with their own internal biological clocks. The insect knows what time of day it is and thus it knows where the Sun is in the sky and can use this to sense compass bearing, so it knows whether it is flying North, East, South or West.

How does an Electroretinogram Work?

By measuring the voltage generated across the insect eye in response to a pulse of light, the following type of trace can be seen on an oscilloscope:

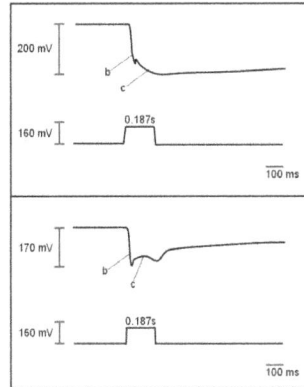

Two examples of an electroretinogram. In each case the square pulse indicates when the flash of light was administered (with time along the bottom horizontal axis) and the trace above each stimulus shows the response of the insect eye. The response of the eye consists of a rapid change in voltage (called the b wave) followed by a slower, longer-lasting change in voltage (called the c wave). Probably the b wave is from the eye itself, and the c wave from the underlying nervous tissue in the optic ganglion of the brain. One of the traces is for a dark-adapted insect eye, the other for a room-light adapted insect eye, can you tell which is which? The top trace is for the dark-adapted eye, which is why the response is larger - the eye has become more sensitive to light (the response is recorded in milliVolts, mV, or thousandths of a Volt and is due to electricity generated by the eye as it encodes light energy into electrical energy). Thus, a bigger response indicates higher sensitivity. The above traces were obtained with white light beam pulses of 0.187 seconds duration. The light was passed through heat filters to remove the heat (so the insect does not respond to the heat as well). Now, by using narrow band-pass filters, it is possible to produce a beam of light of a specified colour. For example, a red filter might let light through at 600 nm (and a bit either side) whilst a blue filter may let through light only at around 450 nm. Thus, one can measure the response of the eye to different colours of light. The responses have been standardised by testing the eye only in the dark-adapted state. The graph below shows the spectrum obtained for female onion flies.

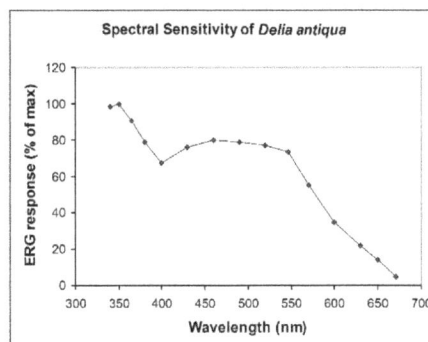

Again we have a peak in the ultraviolet - a very strong peak. Onion flies fly more than Aleochara and so probably need to use the Solar compass more often and perhaps more precisely. Being a rove beetle, Aleochara spends a lot of its time on or under the ground in burrows that it digs, folding its wings under its elytra for protection. Aleochara will only fly in direct sunlight. The onion fly also has good sensitivity in the green part of the spectrum, but it has greater sensitivity to blue light. Again this may indicate a strong flier and as this insect feeds on onion plants, a high sensitivity to the blue-green leaves of onions perhaps helps it to spot them more easily (though odours will also be very important).

So, these insects may at least be dichromatic (with UV and green sensors), possibly trichromatic (with UV, blue and green). However, this data is not enough to prove that they have colour vision. Although the eye has the necessary sensors, we do not know whether the brain interprets colours as brightness or as colour. Behavioural experiments, similar to those done on bees, might be able to answer that. Some moths have peaks in the UV, blue and green and also in the red or infrared and so may be tetrachromatic.

Insect Behavior

It is defined as the way that organisms respond to their environment and to internal signals. In addition to many basic behaviors that are shared by other invertebrates, such as mating, there are some insect species that are capable of more advanced forms of interaction with each other and with their environments. There are two types of behavior that can be observed in organisms: innate behavior and learned behavior. Innate behavior is genetically encoded. Flight and mating habits are considered innate behaviors. You have probably seen a clear example of innate insect behavior called the dorsal light reaction. Flying insects will sense the direction of light coming from the sun and fly in a way that keeps the sun overhead, or on their dorsal side. This is a means for the insect to maintain a flight plan that is parallel to the ground. You may have witnessed that this innate behavior is not so helpful when a moth encounters an artificial light source and flies in continuous circles around it to keep the light on its dorsal surface. At times, the moth is not able to fly away from that light source and, in essence, becomes trapped. Another insect behavior that you have probably experienced firsthand is sound production. Some insects produce sounds in order to communicate. An example is the chirping of crickets that often lulls us to sleep on warm summer nights.

Learned behaviors are those that are not encoded genetically and are not present in the organism at birth. They are obtained through life experiences, and they can change or improve over time. Learned behaviors require acute sensing of environmental signals and a fairly complicated network of nerve cell connections for transmitting those signals in order to process the information and modify or initiate a behavior. Insects are capable of this level of behavior. In order to forage for food and return to the same food source repeatedly, they use a type of learning called associative learning. Associative learning is when separate ideas or environmental stimuli are connected to each other. For example, the location of a food source can be associated with a series of visual cues seen on the way to the source. In this way, the organism learns that if it follows a path that includes all of those visual cues, it will again find the food source. There are examples of this type of learning that can be tested experimentally. For example, honeybees can be taught to obtain their food from a particular source based on color cues, even when the location of that source changes. They can learn that their food (sugar water) is located on a yellow dish next to a blue dish containing only water. If the dish positions are switched, the bees remember which color has the food, and they seek that dish.

There are several groups of insects that have evolved complex social networks that involve elaborate patterns of communication between individuals within the community. These species are called social insects.

Social Insects and Communication

Insects and other organisms that live together in well-organized and tightly integrated colonies are called eusocial animals. Eusocial insects include species of ants, termites, bees, and wasps. Some colonies can include millions of individual animals. These are two of the major features of eusocial insects:

- Division of reproductive labor.

- Cooperative care of the young members of the colony.

Social ants are a good example of the division of reproductive labor. The individual ants within a colony divide up into three major groups:

- Fertile females (queens).

- Infertile (sterile) females (workers).

- Fertile males (drones).

Fertile males and queens carry out the reproductive activities of the colony, while workers focus on obtaining food as well as building and maintaining the nest or hive. In some social insect colonies there are other specialized individuals; nurses, for example, feed and care for young larvae. Often times, some individuals in a colony will form a defensive army with the main purpose of defending the nest.

A Termite Nest: This cathedral-like structure is the nest of a huge colony of termites in Australia. In fact, it is the world's largest known termite nest. It towers 7.5 meters (25 feet) above the ground and houses millions of termites.

Communication between members of a colony can take several different forms. Ants generally communicate using pheromones. Pheromones are hormones that are released by one individual to be sensed and responded to by another individual. You may have noticed how a group of ants making their way to a crumb on the ground will generally travel lined up one in front of the other. This is because they are sensing and following a trail of pheromones that was laid down by an earlier ant that discovered the crumb. On its way back to the nest, this little pioneer (called a forager) left a trail of pheromone for other colony members to follow.

Honeybees have evolved a fascinating form of communication using body movement. These insects perform an elaborate dance called the "waggle dance" to tell other colony members where to find a source of food. The angle of the dance indicates the particular direction of the food source relative to the sun, and the length of the dance correlates with how far away the food is, This is considered a form of abstract symbol communication, meaning that they use a behavior to represent information (in this case a location) about something in the environment.

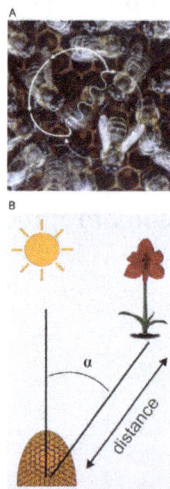

Bee communication. Some bees can tell each other the location of a food source by performing a "waggle dance." The dance is outlined in this diagram. The angle (alpha) of the "waggle," represented by the wavy section of the bee's path, from the sun indicates the direction of the food source. The length of the "waggle dance" indicates how far away the food source is.

In addition to the ability to learn, some species of social insects have also been shown to teach behaviors to other individuals. An example of this is seen in one ant species where the forager ant (the one who ventures out to find food) will actually take the time to lead a nest-mate to a new food source, in a way teaching the nest-mate how to forage.

Altruism is another feature of many social insects. Altruism is the act of self-sacrifice for the benefit of others. A worker bee, for example, forfeits her own potential for reproduction in order to obtain food and provide shelter for the benefit of the queen bee. This allows the queen bee to focus on the reproduction of her genetic material at the cost of the worker bee reproducing her own genetic material.

Most of us think ants are just pests. But not Brian Fisher. Known as "The Ant Guy," he's on a mission to show the world just how important and amazing these little creatures are. In the process, he hopes to catalog all of the world's 30,000 ant species before they become casualties of habitat loss.

Ladybugs, also known as ladybird beetles, have a life cycle of four to six weeks. In one year, as many as six generations of ladybird beetles may hatch. In the spring, each adult female lays up to 300 eggs in small clusters on plants where aphids are present. After a week the wingless larvae hatch. Both the ladybird beetle larvae and adults are active predators, eating only aphids, scales, mites, and other plant-eating insects. The ladybugs live on the vegetation where their prey is found, which includes roses, oleander, milkweed, and broccoli. Adult ladybugs don't taste very good. A bird careless enough to try to eat one will not swallow it.

By late May to early June, when the larvae have depleted their food supply, the adults migrate to the mountains. There, they eat mainly pollen. The ladybugs gain fat from eating the pollen, and this tides them over during their nine-month hibernation. Thousands of adults hibernate overwinter in tight clusters, called aggregates, under fallen leaves and ground litter near streams. In the clear, warmer days of early spring, the ladybugs break up the aggregates and begin several days of mating.

Innate Behavior

Innate behavior is genetically programmed. Individuals inherit a suite of behaviors (often called an ethogram) just as they inherit physical traits such as body color and wing venation. In general, innate behaviors will always be:

- Heritable — encoded in DNA and passed from generation to generation.

- Intrinsic — present in animals raised in isolation from others.

- Stereotypic — performed in the same way each time by each individual.

- Inflexible — not modified by development or experience.

- Consummate — fully developed or expressed at first performance.

Since innate behavior is encoded in DNA, it is subject to genetic change through mutation, recombination, and natural selection. Just like physical traits, innate behaviors are phylogenetic adaptations that have an evolutionary history.

Comparative study of similar species often sheds light on the selective pressures that drive evolutionary changes in behavior. It may also help explain the origin of some very unusual behavior. One species of dance fly, for example, has a courtship ritual in which a male gives a ball of silk to a female. She unravels the ball while he mates with her. By itself, this curious behavior seems truely bizarre. But a study of courtship in other dance flies reveals that males use a nuptial gift as a way to divert a female's aggressive behavior long enough for insemination to occur. In "primitive" species, the nuptial gift is an item of prey that the female consumes during copulation. In more "advanced" species, males wrap the prey insilk, thus buying a little extra time for copulation. In the species where males offer just a ball of silk, they are exploiting the female's innate response to the stimulus of a nuptial gift. Just another example of "selfish genes" at work.

Just because an insect's behavior is innate does not necessarily mean it is simple. Over time, natural selection can lead to surprisingly intricate and sophisticated behavior such as the dance language of honey bees or the courtship rituals of dance flies. These behaviors may appear purposeful and intelligent, but they are merely the product of millions of years of genetic refinement through natural selection.

Dance flies mating: The female is consuming a nuptial gift.

In general, innate behaviors are viewed as "programmed" responses to external stimuli. They usually fit into one of the following categories:

1. Reflex: The most basic unit of innate behavior is a simple reflex arc. This is a neural pathway that may involve as few as two neurons: a sensory neuron detects a stimulus and is linked with a motor neuron that sets off a response in an effector cell (such as a muscle or a gland cell). More commonly, reflex arcs also include an association neuron spliced between the sensory and motor neurons. The association neuron also synapses with other neurons to relay information to the brain and other parts of the body. Most insects have simple "startle" reflexes triggered by small disturbances as well as more comprehensive "escape" reflexes triggered by larger disturbances.

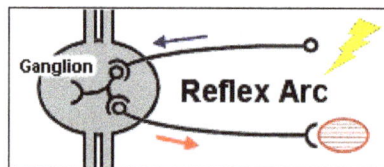

Orientation Behaviors are coordinated movements (walking, flying, swimming, etc.) that occur in response to an external stimulus. These behaviors have adaptive value for survival by helping the insect locate (or avoid) the source of a stimulus. Orientation behaviors can be viewed as elements in a neural hierarchy. The simplest behaviors involve input from only a single sensory receptor whereas more advanced behaviors require bilateral input from a pair of receptors.

2. Kinesis is a change in the speed of movement (orthokinesis) or a change in the rate of turning (klinokinesis) which is directly proportional to the intensity of a stimulus. Input from only a single sensory receptor is necessary. A kinesis is non-directed orientation, that is, the animal exhibits a "random walk". The change in speed or rate of turning increases the probability of locating the stimulus but does not guarantee it.

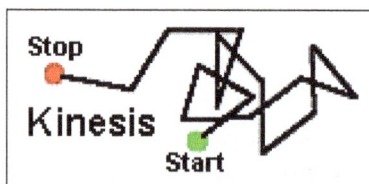

Taxis is a movement directly toward (positive) or away from (negative) a stimulus. A klinotaxis involves side-to-side motions of the head or body with successive comparison of stimulus intensity as the animal moves forward. A tropotaxis requires bilateral input from paired sensory receptors such that the signal is equalized in both receptors. Stimulus intensity increases with movement toward the source and decreases with movement away from the source. A prefix may also be used to designate the type of stimulus involved (i.e. phototaxis=light; geotaxis=gravity; thigmotaxis=-contact with other objects.

The dorsal light reaction is a special case (telotaxis) in which movement occurs at a constant 90° angle to a light source. By keeping the sun directly overhead, a flying or swimming insect can insure that it travels parallel to the ground (or water surface). A diving beetle, for example, can be fooled into swimming upside down in an aquarium that is lit from below. The dorsal light reaction also explains why moths tend to circle a street lamp at night.

The light compass reaction is another special case (menotaxis) in which insects (honey bees, for example) fly away from the nest site at a fixed angle (x°) to the sun and return at the supplementary angle (180-x°) — all without ever taking a class in geometry.

3. Fixed Action Pattern (FAP) is a sequence of coordinated movements that are performed together as a "unit" without interruption. Each FAP is triggered by a unique stimulus variously known as a sign stimulus, a key stimulus, or a releaser. A praying mantis striking at prey is a typical example. The releaser for this FAP is any movement by a small (prey-sized) object within striking distance. Once initiated, the mantis cannot change direction in mid-strike or abort the mission if the prey escapes. Other common examples of FAPs include courtship displays, hunting or food gathering, nest-building activities, and attack or escape movements. Unlike simple reflexes, FAPs may involve a whole-body response and often require a threshold level of internal readiness (drive).

A fixed action pattern is rarely triggered by the "big picture" (Gestalt) in an environmental context. Instead, the sign stimulus is usually a highly specific signal that is consistently encountered at an appropriate time. Thus a male fruit fly will perform a courtship display for a pheromone-impregnated cork even though the cork doesn't look, taste, feel, or act like a female fruit fly. Sometimes, it is possible to find or create a supernormal stimulus, essentially an exaggerated signal that produces a more vigorous or more sustained response than a normal releaser. Apple growers, for example, hang large red spheres coated with stick-um in their orchards to catch adults of the apple maggot (Rhagoletis pomonella). The red spheres are a key stimulus for oviposition, and to female apple maggots, size matters.

Performing one FAP may lead an insect to encounter the releaser for a second FAP and that in turn may lead to the releaser for a third FAP, etc. This type of behavioral cascade is common in insects. Niko Tinbergen, one of the "fathers" of modern ethology, demonstrated that hunting behavior in a predatory wasp proceeds through a stepwise series of three FAPs. The first releaser is visual: movement of a prey-sized object triggers down-wind pursuit of the prey. At that point, prey odor is a releaser that triggers catching behavior. Finally, tactile cues from the prey release stinging and egg laying behavior.

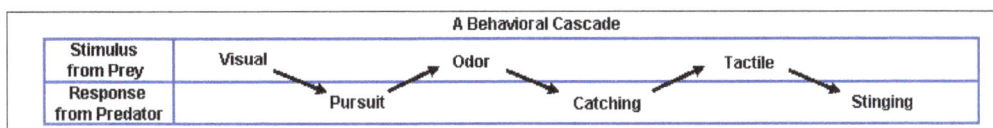

A Behavioral Cascade			
Stimulus from Prey	Visual	Odor	Tactile
Response from Predator	Pursuit	Catching	Stinging

When a fixed action pattern fulfills or satisfies a physiological drive, it may be known as a consummatory act. Taking a blood meal, for example, satisfies a mosquito's hunger drive. Any behavior that increases an individual's probability of encountering the releaser for a consummatory act is often called an appetative behavior. Thus, chirping by a cricket may be regarded as an appetative behavior because it increases the chances of finding a mate and satisfying the sex drive.

Learned Behavior

Learning can be defined as a persistent change in behavior that occurs as a result of experience. Since a newborn nymph or larva has no prior experience, its first behaviors will be entirely innate. Each individual starts life with a "clean slate": it acquires new skills and knowledge through trial and error, observation of other individuals, or memory of past events. In general, learned behaviors will always be:

- Nonheritable — acquired only through observation or experience.

- Extrinsic — absent in animals raised in isolation from others.

- Permutable — pattern or sequence may change over time.

- Adaptable — capable of modification to suit changing conditions.

- Progressive — subject to improvement or refinement through practice.

Although insects have relatively simple nervous systems and are not able to master college-level physics, they have demonstrated the ability to "learn" in each of the following ways:

- Habituation is learning to "ignore" stimuli that are unimportant, irrelevant, or repetitive. For example, a puff of air on the cerci of a cockroach will cause the animal to scamper away. But repeating the same stimulus over and over will lead to a decrease in the response and eventually to no response at all. In some insect populations, widespread use of sex pheromone will disrupt mating behavior. By making everything in the world smell like a virgin female, males become habituated to the odor and stop responding to the signal. If a female cannot attract a mate, she will not produce any offspring.

- Classical Conditioning is learning to associate one stimulus with another, unrelbeesated stimulus. Honey bees, for example, learn to associate floral colors and fragrances with the presence of nectar. They can be "trained" to collect sugar water from colored dishes on a feeding table. If a blue dish with pure water sits next to a yellow dish with sugar water, worker bees will quickly learn to associate "yellow" with "food" (even if the dishes are moved around). When solutions in the two dishes are suddenly swapped (sugar to blue and water to yellow), the bees will ignore blue and continue to forage at yellow until they eventually "learn" (by trial and error) to look for the blue dish.

- Instrumental Learning depends on the animal's ability to remember the outcome of past events and modify future behavior accordingly. Good consequences (positive feedback) reinforce the behavior and increase its likelihood of occurrence in the future. Bad consequences (negative feedback) have the opposite effect. Cockroaches learning to run through a simple maze to find food is a simple example of instrumental learning (also known as operant conditioning).

- Latent Learning involves memory of patterns or events when there is no apparent reward or punishment associated with the behavior. A sand wasp, for example, learns the location of her nest site by taking a short reconnaissance flight each time she leaves the nest. She remembers the pattern of surrounding landmarks to help her find the nest when she returns. Likewise, worker ants can remember a series of landmarks along a trail and follow them (in reverse order) back home to the nest site. Honey bees also show latent learning when they follow the waggle dance of a forager and then use that information to find the reported nectar source.

Honey bees at a feeding station

- Imprinting is a special case of programmed learning that occurs early in life and only within a short time-window known as the "critical period". During this brief interval, the animal acquires an indelible memory of certain salient stimuli in its "home" environment (taste of the host plant, smell of the nest site, etc.). This memory is retained throughout life and recalled later when needed. Fruit fly larvae, for example, will imprint on the taste and smell of their food. If reared on a diet that contains apple extract, adult females will show a strong preference for apples when they eventually search for a place to lay their own eggs. Not just any stimulus will do. Imprinting is apparently regulated by an innate "neural template" that restricts what can be remembered.

References

- Insect-senses: daviddarling.info, Retrieved 11 May, 2019

- Mechanoreception-touch-sensillar-structure-insects, insects: what-when-how.com, Retrieved 20 July, 2019

- Insect-thermohygroreceptors, BioTech: cronodon.com, Retrieved 19 March, 2019

- Photoreceptors, senses: cals.ncsu.edu, Retrieved 9 January, 2019

- Chemoreception-insects, insects: what-when-how.com, Retrieved 29 April, 2019

- Insect-Vision, BioTech: cronodon.com, Retrieved 12 June, 2019

- Insect-Behavior-Advanced, insect-behavior, biology: ck12.org, Retrieved 2 February, 2019

- Innate-behavior, elements-of-behavior: cals.ncsu.edu, Retrieved 10 August, 2019

- Learned-behavior, elements-of-behavior: cals.ncsu.edu, Retrieved 17 April, 2019

Chapter 5

Defense Mechanisms in Insects

Insects are threatened by a broad range of predators, such as birds, amphibians and carnivorous plants. The different ways in which insects defend themselves are classified as mimicry, mechanical defence and chemical defence. The chapter closely examines these key concepts of insect defence to provide an extensive understanding of the subject.

For many insects, a quick escape by running or flying is the primary mode of defense. A cockroach, for example, has mechanoreceptive hairs (setae) on the cerci that are sensitive enough to detect the change in air pressure that precedes a fast moving object (like your foot). Nerve impulses from these receptors travel through giant neurons to thoracic ganglia at speeds up to 3 meters per second, triggering an evasive response by the legs in less than 50 milliseconds. House flies have a similar reaction time when you try to swat them. They leap into the air and begin flapping their wings 30-50 milliseconds after sensing a threat.

Tiger moths (family Arctiidae) can detect ultrasonic echolocation by bats. At low intensity, they fly away from the bat, but if the bat's call increases to a certain threshold they quickly drop from the air in an evasive, looping dive. Other alarm reactions may be less dramatic, but just as effective: Madagascar cockroaches hiss when disturbed; cuckoo wasps curl up into hard, rigid balls; tortoise beetles have strong adhesive pads on their tarsi and hold themselves tight and flat against a leaf or stem. Other insects simply "play dead" (thanatosis) — they release their grip on the substrate and fall to the ground where they are hard to find as long as they remain motionless.

An insect's hard exoskeleton may serve as an effective defense against some predators and parasites. Large weevils are notorious for their hard bodies – as you may discover for yourself the first time you bend an insect pin trying to push it through the thorax. Most diving beetles are hard, slick, and streamlined; even if you can catch them, they will often squirm out of your grip.

Spines, bristles, and hairs may be effective mechanical deterrents against predators and parasites. A mouthful of hair can be an unpleasant experience for a predator and parasitic flies or wasps may have a hard time getting close enough to the insect's body to lay their eggs. Some caterpillars incorporate body hairs into the silk of their cocoon as an additional defense against predation.

Some insects have a "fracture line" in each appendage (often between the trochanter and the femur) that allows a leg to break off easily if it is caught in the grasp of a predator. This phenomenon, called autotomy, is most common in crane flies, walkingsticks, grasshoppers, and other long-legged insects. In most cases, sacrificing a limb in this manner creates only a minor disability. In fact, walkingsticks (especially young nymphs) may regenerate all or part of a missing appendage over the course of several molts.

Chemical Defenses

Many insects are equipped to wage chemical warfare against their enemies. In some cases, they manufacture their own toxic or distasteful compounds. In other cases, the chemicals are acquired from host plants and sequestered in the hemolymph or body tissues. When threatened or disturbed, the noxious compounds may be released onto the surface of the body as a glandular ooze, into the air as a repellent volatile, or aimed as a spray directly at the offending target.

Defensive chemicals typically work in one of four ways:

- Repellency — A foul smell or a bad taste is often enough to discourage a potential predator. Stink bugs, for example, have specialized exocrine glands located in the thorax or abdomen that produce foul-smelling hydrocarbons. These chemicals accumulate in a small reservoir adjacent to the gland and are released onto the body surface only as needed. The larvae of certain swallowtail butterflies have eversible glands, called osmeteria, located just behind the head. When a caterpillar is disturbed, it rears up, everts the osmeteria to release a repellent volatile, and waves its body back and forth to ward off intruders.

- Induce cleaning — Irritant compounds often induce cleaning behavior by a predator, giving the prey time to escape. Some blister beetles (family Meloidae) produce cantharidin, a strong irritant and blistering agent that circulates in their hemolymph. Droplets of this blood ooze from the beetle's leg joints when it is disturbed or threatened — an adaptation known as reflex bleeding. Irritant sprays are produced by some termites, cockroaches, earwigs, stick insects, and beetles. The notorious bombardier beetles store chemical precursors for an explosive reaction mixture in specialized glands. When threatened, these precursors are mixed together to produce a forceful discharge of boiling hot benzoquinone and water vapor (steam).

- Adhesion — Sticky compounds that harden like glue to incapacitate an attacker. Several species of cockroach guard their backsides with a slimy anal secretion that quickly cripples any worker ants that launch an attack. Similarly, members of the soldier caste in nasute termites have nozzle-like heads equipped with a defensive gland that can shoot a cocktail of defensive chemicals at intruders. The compounds, which are both irritating and immobilizing, have been shown to be highly effective against ants, spiders, centipedes, and other predatory arthropods.

- Cause pain or discomfort — Saddleback caterpillars, larvae of the io moth, and various other Lepidopteran larvae have hollow body hairs that contain a painful irritant. Simply brushing against these urticating hairs will cause them to break and release their contents onto your skin. The consequence is an intense burning sensation that may last for several hours. Many ants, bees, and wasps (the aculeate Hymenoptera) deliver venom to their enemies by means of a formidable stinger (modified ovipositor). The venom is a complex mixture of proteins and amino acids that not only induces intense pain but may also trigger an allergic reaction in the victim.

Swallowtail larva
with osmeteria

Protective Coloration

Biologists recognize that there is usually an underlying rationale for the great diversity of shapes and colors found in the insect world. We may not know why a particular species has parallel ridges on the pronotum or black spots on the wings, but we can be reasonably certain that this shape or color has contributed in some way, however small, to the overall fitness of the species. It is obvious that at least some of the colors and patterns serve a defensive function by offering a degree of protection from predators and parasites. These patterns, collectively known as protective coloration, fall into four broad categories:

1. Crypsis — Insects that blend in with their surroundings often manage to escape detection by predators and parasites. This tactic, called cryptic coloration, involves not only matching the colors of the background but also disrupting the outline of the body, eliminating reflective highlights from smooth body surfaces, and avoiding sudden movements that might betray location. Obviously, this tactic loses much of its effectiveness if an insect moves from one type of habitat to another. Well-camouflaged insects usually stay close to home or make only short trips and return quickly to the shelter of their protective cover. Many ground-dwelling grasshoppers and katydids, for example, have colors of mottled gray and brown that help them "disappear" against a background of dried leaves or gravel. On the other hand, closely related species that live in foliage are usually a shade of green that matches the surrounding leaves. The larvae of some lacewings improve their camouflage by attaching bits of moss or lichen from their environment onto the dorsal side of their body.

2. Mimesis — Some insects "hide in plain sight" by resembling other objects in the environment. A thorn could really be a treehopper; a small twig might be a walkingstick, an assassin bug, or the caterpillar of a geometrid moth; and sometimes a dead leaf turns out to be a katydid, a moth, or even a butterfly. This "mimicry" of natural objects is often known as mimesis. It goes far beyond imitation of plant parts:

    •   Some swallowtail larvae resemble bird droppings, others have falseeyespots on the thorax that create a convincing imitation of a snake's head.

    •   The likeness of a caterpillar can be found on the outer edge of many lepidopteran wings, perhaps serving to fool predatory birds that may peck at the wing margin instead of the butterfly's body.

- Many butterflies and moths have eyespots on the wings that emulate the face of an owl or some other large animal.

- Slug caterpillars and hag moth larvae look like hair balls or small furry mammals.

3. Warning Colors — Insects that have an active means of defense (like a sting or a repellent spray) frequently display bright colors or contrasting patterns that tend to attract attention. These visually conspicuous insects illustrate aposematic coloration, A predator quickly learns to associate the distinctive coloration with an "unpleasant" outcome, and one such encounter is usually enough to insure avoidance of that prey in the future.

4. Mimicry — If a distinctive visual appearance is sufficient to protect an unpalatable insect from predation, then it stands to reason that other insects might also avoid predation by adopting a similar appearance. This ploy, essentially a form of "false advertising", was first recognized and described by Henry W. Bates in 1861. Today, it is commonly known as Batesian mimicry. Viceroy butterflies (mostly palatable to birds) are largely protected from predation because they resemble monarch butterflies (very distasteful). Many species of bee flies, flower flies, robber flies, and clear-winged moths are similarly protected because they mimic the appearance (and often the behavior) of stinging bees and wasps. Batesian mimicry is usually a successful strategy as long as the model and mimic are found in the same location, the mimic's population size is smaller than that of the model, and predators associate the model's appearance with an unpleasant effect.In 1879, Fritz Müller recognized that two or more distasteful species often share the same aposematic color patterns. Many species of wasps, for example, have alternating bands of black and yellow on the abdomen. This defensive tactic, commonly known as Müllerian mimicry, benefits all members of the group because it spreads the liability for "educating the predator" over more than one species. In fact, as the number of species in a Müllerian complex increases, there is a greater selective advantage for each individual species. Mimicry has been carried to extremes in some tropical Lepidoptera where both related and unrelated species resemble each other in size, shape, color, and wing pattern. Collectively, these butterflies (and sometimes moths) form mimicry rings that may include both palatable and unpalatable species. In South America, for example, longwing butterflies (Family Nymphalidae) form a mimicry ring that includes at least twelve different species (including one moth).

Butterfly mimicy ring

Although natural selection favors individuals in a population with the best camouflage or mimicry, it also favors the predator or parasite with the best prey-finding acumen. As a result of these competing interests, coevolution between predator and prey populations inevitably leads to an on-going escalation of offensive and defensive measures — a scenario that Leigh Van Valen of Chicago University describes as an evolutionary "arms race".

In order to survive in the arms race, both predator and prey must constantly evolve in response to the other's changes. Failure to "keep up" concedes a competitive advantage to the opponent and may lead to extinction.

Insect Defense Strategies

Acting

Things that eat other things tend to quickly lose interest in dead prey, so some insects that employ the strategy of playing dead (thanatosis) can often escape unharmed. Threatened insects simply let go of whatever they happen to be hanging on to and drop, motionless, to the ground where they put on the performance of a lifetime. Certain caterpillars, ladybugs, many beetles, weevils, robber flies, and giant water bugs all employ this technique.

Acid and Burning Agents

Some bugs release irritants so awful that it automatically makes "get it off, get it off," the only thing a predator thinks. It's not the most romantic defense mechanism, but it gives the insect time to escape. Some blister beetles produce a blistering agent. Droplets of this will ooze from the beetle's leg joints when it is disturbed or threatened - an adaptation known as reflex bleeding. Some termites, cockroaches, earwigs, stick insects, and beetles will literally spray acid at attackers. The (fantastic) bombardier beetle stores chemicals in specialized glands, and when threatened, mixes them to-gether to produce a forceful discharge of boiling hot quinone and water vapor (steam).

Some caterpillars have hollow body hairs that contain a painful irritant. Simply brushing against the pretty, soft-looking fluff will cause them to break and get all over your skin, resulting in an in-tense burning that may last for several hours. Many ants, bees, and wasps deliver venom to their enemies by means of a formidable stinger (modified ovipositor). The venom is a complex mixture of proteins and amino acids that not only induces intense pain but may also trigger an allergic reaction in the victim.

Speed and Movement

Some insects don't have fancy weaponry or acting chops and have to rely on speed to get away. For many insects, a quick escape by running or flying is the primary mode of defense.

House flies have an insanely fast reaction time when you try to swat them. They can fly away 30-50 milliseconds after sensing a threat. And a cockroach? Those nasty things have tiny, super-sensitive hairs that are acute enough to detect the change in air pressure that might occur right before you try to step on them. It can react in less than 50 milliseconds. Just try and hit them with a newspaper.

Armor and Spines

Plenty of insects simply rely on their exoskeleton to avoid being eaten.

Large weevils have ridiculously strong armor and can actually bend insect pins when bug collectors try to add them to their boards. The Ironclad Beetle can literally be stomped on, and take zero damage. These guys don't just bend those insect pins, you need a drill to get through their shell. Then there are spines, bristles, and hairs that work pretty damn good in discouraging things that might be hungry. Nobody wants a mouthful of needles, and the hairier insects make for an unpleasant dining experience as well. Plus, insects like super-hairy caterpillars make it super difficult for parasitic flies or wasps to even get close enough to the insect's body to lay their eggs.

Mimicry

False advertising. Basically, these insects are piggybacking on a good thing that other insects developed the hard way. For example, Monarch butterflies have a free pass because they taste, really, really bad to birds. So the Viceroy butterfly thought to itself - ok, cool. And went off and bought the same dress. Viceroys look exactly like Monarchs, despite the fact that they taste just fine. Many species of wasps, as we all know, have alternating bands of black and yellow on the abdomen. The robber fly has developed those same stripes in order to pass as something dangerous. There are spiders that mimic ants, flies that mimic bees, and even entire groups of insects display mass behaviours like beetle larvae that co-operate to mimic bees. Find something that's working for other bugs and copy it.

Camouflage

Unlike the mimics, these insects have worked hard to blend in on their own terms.

Insects that blend in with their surroundings often manage to escape detection by predators and parasites. Birds can't eat what they don't see in the first place. This tactic involves not only matching the colors but also disrupting the outline of the body and eliminating reflective highlights. Also, holding very still... because once you move, obviously you won't look so much like whatever it is you're standing against. These guys usually stay close to home or make only short trips out and back - since their coloration tends to be area-specific. Lots of grasshoppers and katydids have colors that work great against a background of dried leaves or gravel, but not so well against green leaves. Conversely, other species that live in those leaves are usually a shade of green that matches.

Mimesis

This is different than mimicry or camouflage, though it uses the same principle. Some insects "hide in plain sight" by resembling objects in their environment. A thorn could really be a treehopper; a twig might be a walkingstick, an assassin bug, or a caterpillar; and sometimes a dead leaf turns out to be a katydid, a moth, or even a butterfly. Some caterpillars resemble bird droppings, and others have false eyespots on their wings or body to create an imitation of a predator's head. Often, these guys are the coolest-looking... the details in their appearance astonishing in their accuracy and creativity.

Repellency

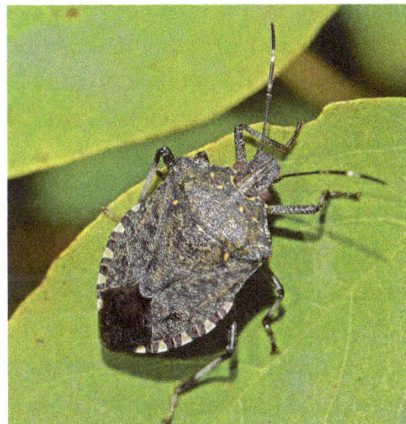

If there is one thing most of us have in common, it's a distaste for foul smells. And the really bad ones can be enough to make you recoil. Ever been at the epicenter of a skunk attack? It's like someone is burning tires directly in your NOSE. Stink bugs have special glands that produce a foul-smelling reek. The caterpillar form of some swallowtail butterflies have glands just behind their heads, that, when disturbed, will rear up and release a terrible stench. Darkling beetles will raise their big, black butt in warning when they are threatened, and if you don't pay attention to the warning - will expel acrid, foul-smelling fluid.

Adhesion

When stink and burning isn't enough, some bugs will hit their attackers with sticky compounds that harden like glue and incapacitate. Some kinds of cockroaches guard their backsides with a slimy anal secretion (those are three words that are just terrible together) that cripples any ants that launch an attack. And there are types of soldier termites that have nozzle-like heads that can spays sticky, immobilizing toxic fluids at attackers as varied as ants, spiders, centipedes, and other predatory arthropods.

Mimicry

Mimicry is the adaptive resemblance in signal between several species in a locality. The most spectacular and intriguing cases are those of accurate resemblance between distantly related animals, such as spiders mimicking ants. Closely related species can also benefit from mutual resemblance, in which case mimicry results from selection against signal divergence.

The vast majority of the hundreds of thousands of insect species are described and identifiable on the basis of morphological characters. This bewildering diversity is, however, ordered because species share characters with their relatives—and one of the taxono-mist's tasks is indeed to recognize, among the shared and divergent characters, a sign of the relatedness of the taxa. Nevertheless, some distantly related species may share a common morphology. Such resemblance may be the result of evolutionary convergence, that is, parallel lifestyles leading to the selection of similar morphological structures; in this case, resemblance per se is not under selection. On the contrary, when a character is taken as a signal between individuals, one species may benefit from bearing the same signal as the one already used by another species; then selection acts directly to favor increased resemblance.

An Interaction between three Protagonists

The Discovery of Mimicry and the Development of Evolutionary Hypotheses

Mimicry in insects has been a puzzle for entomologists long before the Darwinian concept of natural selection, but the explanations for mimicry are tightly linked to the development of evolutionary thinking. While he was traveling in the Amazon with Alfred Russel Wallace in 1842, British entomologist Henry Walter Bates noted that distantly related butterfly species bore the same wing color pattern. Moreover, these communities of species changed their shared pattern in concert across localities. Among these species were the very abundant Ithomiinae (called Danaoid Heliconiidae then, now a subfamily in the Nymphalidae) and rarer Dismorphiinae (called Leptalidae then, now a subfamily in the Pieridae). Bates, as a pioneer evolutionist (but after Darwin published his On the Origin of Species), developed an adaptive explanation for the resemblance. Hypothesizing that ithomiines were inedible to most predators, he proposed that the edible pierids would benefit from being mistaken for their defended counterparts and would thus be selected to resemble them. Edward B. Poulton later named this kind of mimicry after him as Batesian mimicry, when an edible species mimics a distasteful one.

Bates also realized that some apparently inedible ithomiine species in the genus Napeogenes seemed to mimic other inedible Ithomiinae. He proposed that, in fact, rare species, whatever their palatability, should benefit from resembling defended common species. It was, however, more difficult to understand the resemblance of abundant and distasteful Melinaea, Mechanitis (Ithomiinae), Lycorea (Danainae), and some Heliconius (Heliconiinae) from Peru and Colombia, so he assumed the resemblance was the result of some inorganic or environmental factors. In 1879, German naturalist Fritz Muller was the first to develop a mathematical demonstration that two unpalatable prey could benefit from mutual resemblance. He understood that, if the community of predators had to kill a certain (fixed) number of prey to learn to avoid them, two indistinguishable distasteful species would together suffer this mortality and both reduce their death rate per unit time. Muller actually showed that this benefit was biased in favor of the rarer species, to a factor equal to the square of the ratio of the species' abundance. Therefore, unequal population sizes translate into even more unequal, although still mutual, benefits: Mu-lerian mimicry, thus defined, could be beneficial for both species, and perhaps also for the predators, in contrast to parasitic Batesian mimicry.

Mimicry: An Interaction between Senders and Receivers

Mimicry typically involves at least three protagonists, two senders and one receiver, with the receiver judging the resemblance of the signals from the two senders. Obviously, both the senders and the receiver should be found in the same locality for the mimicry to be possible, although time lags or geographic separation between senders may be plausible if receivers have a long-term associative memory and/or migrate. In a habitat, many senders will converge on the same signal, thereby forming what is called a mimicry complex, or mimicry ring. Signals may involve different sensory modalities, depending on the receiver's sensory ecology: static visual signals (e.g., warning color patterns in butterflies, recognizable body shapes in ants), motion (flight behaviors), acoustic signals (clicking in many arctiid moths), olfactory/chemical signals (pherom-ones or the so-called cuticular hydrocarbon profiles by which social Hymenoptera recognize one another), or tactile signals (used by brood parasites of ants to be allowed to enter their nests). Signaling is indeed often multimodal.

Apparent complications may arise when, for example, one of the senders is also the receiver. For example, a predator may mimic the appearance of its prey when approaching it (aggressive mimicry in some spiders or chemical/tactile mimicry for brood parasites); the prey is thus fooled by the predator via its own conspecific signal.

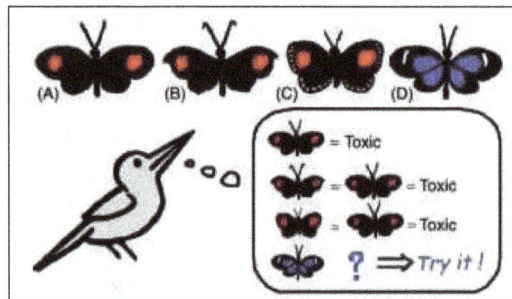

Conditioned predators and signaling prey.

Predators are known to generalize their knowledge of distasteful prey to other resembling prey. Therefore, once predators recognize one prey as distasteful (prey A), other prey may gain from mimicry, whatever be their palatability (prey B and C). If the prey is palatable (prey C), its mimetic gain becomes limited by its abundance in the locality. Finally, a conspicuous prey with a (nonmimetic) pattern new to the predator should suffer higher mortality, making the evolution of diversity in warning color and mimicry a puzzle.

The two senders can also be the same species. This is sometimes the case in chemical-sequestering phytophagous insects when unpalat-ability varies among individuals in the same population (e.g., Danaus gilippus in Florida), leading to so-called automimicry of palatable toward unpalatable individuals in the same species. Similarly, male Hymenoptera do not have the defenses that females have.

It highlights the important evolutionary dynamics that arise from whether receivers are expected to try to discriminate or generalize on the senders' signals or, in other words, from senders sending honest compared to dishonest signals. This should bring into perspective some of the main and still unresolved puzzles in mimicry theory, such as the rise and maintenance of diversity in mimicry signals. Most examples are chosen from the butterfly genera that represent today's best known mimetic organisms, such as Papilio and Heliconius; indeed, our knowledge of the ecology and genetics of mimicry in these genera is unequaled by any other group of insects.

Frequency-dependent Population Processes

Batesian Mimicry and Negative Frequency Dependence

Theory and Consequences: In Batesian mimicry, one of the sender species, the mimic, sends a dishonest signal to deceive the receiver—for example, a predator. It is thought that deception is possible only if the receiver has previously inherited or acquired knowledge about this signal. There is ample evidence that (1) vertebrate predators (birds, lizards) can learn to recognize distasteful prey, (2) they can be deceived by mimicry, and (3) mimics gain from the resemblance. The most famous Batesian mimic is probably the viceroy butterfly, Limenitis archippus . which mimics the monarch D. plexippus, although this relationship is now questioned (because viceroys can be unpalatable).

Hoverflies (Diptera: Syrphidae), diurnal moths (Sesiidae, Sphingidae), striped beetles (Cerambycidae), or crane flies (Tipulidae) are well-known Batesian mimics of wasps and bees.

Batesian mimicry: The day-flying moth Synanthedon tipuliformis (Sesiidae) (top) is a Batesian mimic of stinging wasps in Europe. The resemblance is very accurate, and the moth is very rare compared to its wasp models, so that it is not often observed. Similarly, but in a totally different group, the beetle Clytus arietis (bottom) mimics wasps and is sometimes seen on blossoms. These two examples illustrate how the same general appearance can be achieved by morphological changes of totally different nature in different groups of insects.

Clearly, the efficiency of the deception is directly linked to the probability that predators have knowledge of the prey. It thus depends on the ratio of models and mimics in the population of prey. As in host-parasite systems, the fitness of Batesian ("parasitic") mimics therefore depends negatively on their proportion in the prey community. Negative frequency dependence, the selective advantage to rare forms, is thought to be a strong force favoring and maintaining diversity in many ecological situations in nature. In Batesian mimics, any new (or rare) mutant resembling another protected model will be favored, leading to a balanced polymorphism between the two mimetic forms. Negative frequency dependence also predicts that the local number of Batesian species should be dependent on the abundance of the model(s).

Many, but by no means all, Batesian mimics are indeed polymorphic. Among the most famous is the African swallowtail P. dardanus, which may have three co-occurring forms that mimic different species of the Danainae genus Amauris. Hypolimnas mis-ipppus (Nymphalinae) is another African butterfly that has four forms mimetic of D. chrysippus. In South America, the swallowtail Eurytides lisithous has up to three forms that mimic the co-occurring Parides species (Papilionidae), whereas in Southeast Asia the famous Papilio memnon also mimics three or more different papilionid models. In the Diptera, the Old World hoverflies Volucella bombylans and Merodon equestris are examples of polymorphic species mimicking bumble bees.

Evidence for Negative

Frequency Dependence

Although experimental demonstration that Batesian polymorphisms stem from negative frequency dependence is still lacking, there is a lot of evidence for negative frequency dependence itself. A first line of evidence comes from the observation of patterns of abundance of models and mimics in nature. For example, the North American butterfly Battus philenor is known to be unpalatable to most birds and is believed to act as model for a number of edible mimics in the " black" mimicry ring. In one of them, P. glaucus, females are found as a mimetic and a nonmimetic (male-like) form, and the proportion of the mimetic form tends to be higher where its model B. philenor is more abundant. Similarly, the resemblance of the mimic P. troilus to B. philenor is higher where the latter is abundant. These give an overall pattern of mimics' occurrence consistent with negative frequency dependence. Moreover, field experiments directly showed a strong selective advantage to mimetic vs. nonmi-metic Callosamia promethea day-flying moths, another Batesian mimic of B. philenor.

Experimental approaches give more insight into the mechanisms involved in frequency dependence. In experiments, captive or wild predators can be tested with a variety of artificial or real prey, and the mimic/model proportions can be experimentally changed to explore how it affects the preys' survival. Traditional experiments were carried out in the 1970s with mealworms or pastry baits colored with food dyes, and dipped in quinine to make them distasteful, and exposed to garden birds in suburban Britain. Such experiments do suggest that a rare mimic has an advantage over a common one if the "model" is slightly distasteful, which demonstrates frequency-dependent selection. However, if the "models" were made very distasteful, the advantage of being rare decreased and eventually vanished. Laboratory experiments can also be used to search for evidence of frequency dependence, while avoiding potential confounding effects of field experiments. Experiments with captive great tits as predators showed that the mortality of both mimics and models depended on the frequency of the model and that both models and mimics survived better when mimics were fewer.

These experiments tell us that the intensity of frequency-dependent selection in mimics is highly dependent on the palatability of the models. To see its selective advantage decrease, the palatable mimic must become very common, or the model must be not very distasteful. This suggests there is some kind of effective "equivalence" between relative numbers of prey encountered and their relative levels of toxicity.

Positive Frequency Dependence in Mullerian Mimicry

Theory: The Disadvantage of Rare Forms

Warning signals, or aposematism, evolve because prey bearing signals that predators associate better with unprofitability (e.g., harmful prey) survive better. The evolution of warning signals brings some apparent paradoxes that are not treated in that entry. However, there is plenty of evidence that aposematic prey are easily learned and subsequently avoided by vertebrate predators. Both the warning prey and the learning predator benefit from a correct interpretation of the signal. Under such an "honest signaling" framework, rare or new variants within a prey population should not be recognized as distasteful and should suffer higher predation. This selection against rare forms translates into positive frequency-dependent selection: rare mutants are removed, leading to monomorphism in all populations.

Because predators select only on prey appearance, the selective pressure does not stop at the species boundary: several protected prey species may be selected to use the same warning signal, that is, become Mullerian mimics. Although the phenomenon is not necessarily symmetrical, two or several defended species should all benefit from sharing a warning signal, which reduces their per capita pre-dation rate.

Six butterfly mimicry rings from eastern Peru. The mimicry rings (groups of mimetic species) presented here are dominated by butterflies in the Ithomiinae and occur in the forests around the city of Tarapoto. Following G. W. Beccaloni's nomenclature, these mimicry rings are Tiger, Melanic tiger, Large transparent, Small transparent, Small yellow, and Orange-tip mimicry rings. At least 5 other mimicry rings can be recognized involving Heliconiinae and Ithomiinae in this area, which brings the total to at least 11 mimicry rings for these two butterfly subfamilies. Many more species, not featured here, belong to these mimicry rings, particularly Ithomiines and especially in the Small transparent group. The Tiger mimicry ring involves a lot of species and the size distribution is almost continuous from small to very big. This may be because as more and more Mu-lerian mimics join the mimicry ring, predators might generalize more, and the selection for close resemblance could be somewhat relaxed. Note that some day-flying moths participate in these mimicry rings, probably as Mullerian mimics (they reflex-bleed bitter hemolymph when handled). Butterflies 13-16 and 31 are supposed to be Batesian mimics as they belong to palatable groups within their families. See more species belonging to these mimicry rings in Figures. All butterflies are Nymphalidae: Ithomiinae, except 1-3 (Nymphalidae: Heliconiinae), 14 (Nymphalidae: Melitaeinae), 16 (Nymphalidae: Charaxinae), 15 (Papilionidae), 13 and 31 (Pieridae: Dismorphiinae), 34 (Riodinidae), and 6, 17, and 22 (Arctiidae: Pericopinae). Scientific names: 1, Eueides Isabella; 2, Heliconius pardalinus; 3, H. hecale; 4, Melinaea menophilus; 5, Tithorea harmonia; 6, Chetone histriona; 7, Napeogenes larina; 8, Mechanitis lysimnia; 9, Mec. polymnia; 10, Mec. mazaeus plagifera ssp.; 11, Ceratinia tutia; 12, Hypothyris cantobrica; 13, Dismorphia amphiona; 14, Eresia sp.; 15, Pterourus zagreus; 16, Consul fabius; 17, Chetone histriona; 18, Mel. marsaeus; 19, Hyposcada anchiala; 20, Hypot. mansuetus; 21, Mec. mazaeus deceptus; 22, Notophyson helico-nides; 23, Methona confusa; 24, Godyris zavaleta; 25, Greta andromica; 26, Pseudoscada florula; 27, Notodontid moth; 28, Aeria eurimedia; 29, Ithomia salapia; 30, Scada sp.; 31, Moschoneura sp.; 32, Hypos. illinissa; 33, Hypoleria sarepta; 34, Stalachtis euterpe. Scale bar, 2 cm.

As more and more individuals join in the mimicry ring, the protection given by the signal becomes stronger. Therefore, the direct, and naive, prediction is that all unpalatable prey of a similar size in a habitat should converge into a mimicry ring.

Evidence for the Frequency-dependent Benefits of Mullerian Mimicry

Although comparative and biogeographical studies give strong support to the theory, the first convincing experimental evidence came from pastry-bait experiments with garden birds that tend to attack rare distasteful baits more often than common ones. Recently, laboratory experiments also showed strong selection against new rare warningly colored prey items. However, field evidence with free-living prey is crucial for a validation of these results. In one experiment, J. Mallet reciprocally transplanted Heliconius erato individuals between populations in which H. erato have different wing patterns, thus effectively releasing rare "mutant" and "control" butterflies into the host populations. A strong selective advantage of about 50% was calculated for the commoner form. More recently, to avoid the potential pathology of color patterns being adaptations to local habitat conditions in addition to mimicry, D. D. Kapan used a similar reciprocal release-recapture technique but used polymorphic populations of the butterfly H. cydno. In this species, two morphs coexist but participate in two different mimicry rings that differ in relative abundance in different locations in Ecuador. Life expectancy was 12 days for the locally common forms and only 2 days for the locally uncommon forms. These field data give unequivocal evidence for strong selection against rare forms in these Mullerian species.

Consequences and Challenges

Strong purifying selection now seems well supported by theoretical, comparative, and experimental evidence. To evolve a new pattern, a toxic prey would have to pass an apparently impassable initial disadvantage, survive a transient polymorphism, and win the aposematic competition with alternative warning signals. It is therefore no surprise that most distasteful Mullerian mimics are indeed monomorphic in local populations and that polymorphisms are usually restricted to narrow hybrid zones between color-pattern races. In H. erato, in which two color races abut, frequency-dependent selection maintains a sharp boundary, alternative forms being positively reinforced on either side of a steep cline. Many species join Mu-lerian mimicry rings, which itself represents interspecific evidence for strong frequency-dependent selection.

However, in contrast with such extremely conservative forces, diversity is present at all levels in mimicry. At a macroevolu-tionary level, aposematic and mimetic groups typically undergo rapid mimetic radiations into numerous species and races differing in color pattern, like heliconiine butterflies or pyrrhocorid red bugs. At the community level, many radically different mimicry rings coexist in the same habitat (e.g., five or six coexisting rings just within the Heliconius of Costa Rica, at least seven or eight rings just within the Ithomiinae of the Peruvian Amazon—). At the biogeographical level, many aposematic species show a bewildering diversification in more or less sharply defined mimetic races. Finally, at the population level, several chemically defended species show mimetic polymorphism. For instance, the bumble bee Bombus rufocintus has two mimetic forms in North America, the burnet moth Zygaena ephialtes has two sym-patric forms in Italy, the African monarch D. chrysippus has four main color forms coexisting in large areas in East Africa, and the Amazonian H. numata shows the most astounding polymorphic mimicry with up to 7-10 forms in the Andean foothills. In each of these cases, the different forms closely match the different local mimicry rings.

This rampant diversity does not question the existence of frequency dependence itself, but the details of how purifying selection may or may not prevent the evolution of diversity. It may also question the validity of the two classical categories of protective mimicry (Batesian and Mulle-rian) and the existence of a sharp divide between them along the spectrum of prey palatability. Explaining these unexpected cases is therefore central to our understanding of signal evolution in distasteful insects.

Polymorphic Mullerian mimicry. The Amazonian butterfly H. numata (Nymphalidae: Heliconi-inae—right column) is a Mu-lerian mimic in a variety of tiger-pattern mimicry rings. Each popula-tion (here around the city of Tarapoto in Eastern Peru) is polymorphic and up to seven forms may coexist, each being an exceptionally accurate mimic of species in the genus Melinaea (Nymphali-dae: Ithomiinae—left column). Spatial variation in selection pressure is probably what maintains the polymorphism, by a balance between local selection for mimicry of the commonest Melinaea species and movement of individuals (gene flow) between neighboring localities selected for dif-ferent wing patterns. From top to bottom (left column): Melinaea ludovica ludovica, Mel. sat-evis cydon, Mel. marsaeus mothone, Mel. marsaeus phasiana, Mel. menophilus ssp. nov., Mel. meno-philus hicetas. and Mel. marsaeus mothone. (Right column) H. numata forms silvana, elegans, aurora, arcuella, tarapotensis, timaeus, and bicoloratus. Scale bar, 2cm.

The Palatability Spectrum and Predator Psychology

Models of Mimicry Evolution

Case studies and experiments on mimicry are practically difficult, time consuming, and inform us only on potential processes in particular cases. They are thus not always very informative as to which processes are generally important in the evolution of mimetic diversity. For these reasons, mathematical models simulating mimicry evolution have been widely used. Models of mimicry evolution have been traditionally of two different types: "evolutionary dynamics" models have concentrated on trait evolution in the prey populations, underestimating the effects of the details of predator behavior; " receiver psychology" models have concentrated on the effect of predator cognitive abilities in driving the costs and benefits to mimetic prey, but largely ignored evolution-ary processes in the prey populations, particularly frequency or density dependence. The second category of models are those that apparently pose a threat to the validity of the Batesian/Mullerian distinction, and M. P. Speed even coined the new term " quasi-Batesian mimicry" for the strange, though possibly common, intermediate dynamics that his model highlighted.

The main discrepancies lie in the way predators are thought to respond to prey palatability and density. Speed's models assumed that predators attack a fixed fraction of a prey in a population, irrespective of their total number (linear frequency dependence), and that this fraction depends on the palatability of the species. In a mixture of prey of differing palatability, the resulting fraction killed would be intermediate between the fractions lost in each prey in the absence of mimicry, leading to one prey species benefiting and the other suffering from mimicry. This view, however, leads to the strange prediction that as more mildly unpalatable prey are present, the attacked fraction (per unit time) can increase. In contrast, J. Mallet and M. Joron argued that predators are unlikely to be sensitive to frequency per se and should instead need only to attack a fixed number of prey before learning, making the "attacked fraction" a decreasing function of the total number of prey bearing the pattern. This should lead to a strongly nonlinear, effectively hyperbolic frequency dependence. The attacked fraction (per unit time) should always decrease when the total number (and therefore the density) of unpalatable prey increases, whatever be their relative unpalatability.

The debate is still very much active, and decisive data are surprisingly scarce. In an experiment with pastry baits and wild passerines, Speed showed that the attack fraction of a mimetic pair was indeed intermediate between that of either " species " alone. Furthermore, birds seemed to learn only to a certain extent; that is, they never completely stopped attacking the unpalatable items. Despite some potential problems in the experimental design (especially, the artificially high prey density in a time of scarce prey density), these data remain a puzzle and may hint at more complex learning processes than a pure number-dependent dose response. More decisive evidence came from L. Lindstrom's study, in which novel toxic prey were introduced into a great tit's foraging arena at varying frequencies (= densities in this setting). Although the total number of attacked toxic prey increased with their initial frequency, the attack fraction decreased. Her data support the validity of nonlinear frequency dependence, although the idea of a strictly fixed number of prey killed could be an oversimplification. Absolute numbers of prey attacked may increase with warning signal density, but proportion will inevitably decrease, which should lead to a traditional Batesian-Mullerian distinction. More recently, in an experiment where the total number of signaling prey individuals was allowed to increase with the addition of a mimetic species, thereby approaching more natural conditions for the evolution of mimicry, Speed's group provided long-awaited evidence that mimicry unpalatable prey of varying strength benefited mutually from the mimicry due to an increased density of the signal in the prey population. The mutual benefits may, however, be modulated by the availability of alternative prey to predators in the habitat, as this and theoretical models have suggested. It appears optimal foraging, in which predators tune their discrimination abilities to the availability of prey and their pal-atability distribution, should be taken into account to understand the interplay between density-dependence and the unpalatability spectrum in the evolution of mimicry.

More intriguing prospects have come from the recent discovery by J. Skelhorn that the benefits of mimicry tend to be enhanced when mimics use different types of chemical defenses for protection. Although the mechanisms underlying this are unknown, the potential consequences for speciation and the community ecology of insect species assemblages may be large if selection facilitates the recruitment of distantly related mimics more easily than close relatives.

The Strength of the Selection

Muller's number-dependent model also leads to a prediction that has hitherto been largely over-looked. At low densities, selection should act strongly against any transient polymorphism, but at higher densities, selection quickly becomes weak at intermediate form frequencies. This leads to effective neutrality of polymorphism once it is established in abundant species. Kapan's field experiments, in which H. cydno were released at varying density, showed precisely this trend. Polymorphism could therefore be nonadaptive but very weakly selected against by predators.

Numerical Mimicry and Density-dependent Processes

The studies of J. Allen and his collaborators, and others, show that prey selection by predators can be frequency dependent in palatable, cryptic prey, that is, even in the absence of mimicry of unprofitable prey. This is probably caused, in part, from predators using search images when foraging. For instance, at low densities of a particular kind of (palatable) prey, predators usually prey on the more common form, which corresponds to their search image, imposing a negative frequency dependence. Cryptic prey may be globally numerous in a habitat, but because they are camouflaged, their apparent density to predators is bound to be low. This leads to the diversification of cryptic patterns, and perhaps the selection of plastic (partly environmentally induced) color-pattern genetic control, in prey. In contrast, at high density, predators usually prey on the odd phenotypes preferentially, even among perfectly palatable prey, effectively leading to a positive frequency-dependent selection on morphology.

Gregarious palatable prey that are at locally high density and that presumably rely on predator satiation to escape predation, might then be selected for mutual resemblance. Such a prey might be called warn-ingly colored, whereas the appearance itself is not protective. This idea led to the supposition that several prey species that co-occur at unusually high densities, like mud-puddling butterflies or schooling fishes, might evolve "numerical" or "arithmetic" mimicry by simple frequency-dependent predation unrelated to unprofitability. Prey traits like color, shape, and especially locomotor behavior are therefore thought to be under purifying selection in mixed-species aggregations. This attractive idea remains largely untested in insects, although R. B. Srygley proposed the pair of bright orange butterflies Dryas julia (Heliconiinae) and Marpesia petreus (Nymphalinae) as a potential candidate.

Female-limited Mimicry

Some of the most spectacular and best studied cases of Batesian polymorphism are found in swallowtails, and in some species only the female is mimetic. This peculiar tendency to sex-specific polymorphism seems to be restricted to butterflies (Papilionidae and Pieridae), and virtually no other case of sex-limited mimicry seems to be reported for other insects (except for male-limited mimicry in some moths). Female-limited mimicry was often viewed as a result of negative frequency dependence: if mimicry is restricted to one sex, the effective mimetic population size is only about half that of a nondimorphic species, reducing deleterious effects of parasitism onto the warning signal. But this group-selection argument cannot in itself explain why females tend to become mimetic more often than males and why mechanisms arise that restrict the mimicry to one sex. However, more proximal, individual-selection arguments are not lacking. First, mimicry may be more beneficial to one sex than to the other. For instance, female butterflies have a less agile flight because of egg load and a more "predictable" flight when searching oviposition sites, and they suffer higher rates of attacks by visual predators.

Female-limited mimicry in Perrhybris pyrrha (Pieridae), Eastern Peru.

The female (top) is a Batesian mimic of the tiger-patterned Ithomiines and Helicomiines, while the male (bottom) has retained a typical pierid white coloration. Scale bar, 2 cm.

Second, male wing patterns can be constrained by sexual selection, via either female choice or male-male interactions: males could not evolve Batesian mimicry without losing mating opportunities. In experiments with North American swallowtails (of which only females mimic B. phacile-nor), male P. glaucus painted with the mimetic pattern had a lower mating success than normal yellow males; similarly, painted P. polyxenes males had a lower success in male-male fights and therefore held lower-quality territories around hilltops. In these insects, the wing coloration appears to bear signals directed either to conspecific males or to predators, which creates a potential conflict leading to sex-limited polymorphism. It is interesting to note that Papilio and Eurytides species that mimic Parides (Papilionidae) in South America do not exhibit female-limited mimicry; different modes of sexual selection (e.g., absence of territoriality) may operate in the forest understory habitat. In a different ecological setting, diurnal males of the North American silkmoth C. promethea are exposed to visual predators, and mimicry of B. philenor is limited to males; female Callosamia fly at night and benefit more by crypsis during the day.

Mimicry and the Evolution of Signal Form

Resemblance and Homology

Mimicry can arise as soon as the signal is effectively copied, that is, as soon as superficial resemblance is attained. Therefore, mimics usually bear characters similar to those of their models, but these are often clearly nonhomologous in terms of genes and mechanisms of development. For instance, red spots near the base of the wing in P. memnon mimic the spots on the bodies of their models. The trans-lucency and iridescence of distasteful Ithomiinae clearwing butterflies is mimicked by white raylets in dioptine and pyralid day-flying moths and provide the same impression in motion. Similarly, the black-wing patterning of some flies seems to mimic the superposition of wings over the abdomen in their wasp models. Therefore, mimics from distant phylogenetic groups are certainly under very different functional and developmental constraints to create a mimetic impression. Selection will retain the first characters that suddenly increase overall similarity. The initial step made may therefore strongly influence the route selected to achieve mimicry.

The Genetics of Mimicry: Polymorphisms and Supergene Evolution

The Debate

The genetical study of the evolution of mimicry was first dominated by a debate between gradualists (Fisher) and mutationists (Goldschmidt). Goldschmidt proposed that "systemic" mutations could affect the whole wing pattern of butterflies in one step and that models and mimics, although not using the same genes, were using at least the same developmental pathways. Because this view could not account for the obvious nonhomologies, like those pointed out above, Fisher and others claimed that mimicry was achieved by slow microevolutionary steps and the gradual accumulation of resemblance alleles.

Decisive steps toward a resolution of the debate came principally from the study, by C. Clarke and P. Sheppard in the 1960s, of Batesian butterfly mimics in which color pattern is easy to define and analyze and gene effects are straightforward to identify. Polymorphic mimics, particularly Papilio species, of which different forms could be crossed by breeding experiments (including hand pairing), were particularly useful. It appeared that color pattern is mainly inherited at one or few major loci, affecting the whole pattern. From rare recombinants, it could be shown that these loci were in fact super-genes, that is, arrays of tightly linked small-effect genes. Several additional unlinked "modifier" loci were also shown to increase resemblance via interaction and epistasis with the supergene. Goldschmidt's ideas seemed refuted.

However, although supergenes seem to be a necessary condition for the evolution of polymorphism (otherwise numerous nonmimetic, unfit recombinants would be produced), how they evolve is another issue. Theoretical models suggested that supergenes could not be achieved by simple gradual reduction in recombination. In the absence of spatial variation in selection pressures, tighter linkage cannot evolve by small steps via Fisherian gradual evolution, because good combinations of alleles are immediately broken up by recombination. Instead, gene clusters should preexist the evolution of polymorphism.

The Two-step Hypothesis

These results led to a unifying, now widely accepted two-step mechanism of mimicry evolution: (1) mutations at genes of major effect first allow a phenotypic leap achieving an approximate resemblance to a particular model. (2) Once these mutations have increased in the population, resemblance can be enhanced through the gradual selection of epistatic modifiers. This two-step mechanism is supported by three lines of evidence. First, empirical evidence from butterflies suggests the existence of a small number of major-effect genes and numerous small-effect modifiers. In fact some of these genes of major effect could even include a series of regulatory upstream elements and transcription factors, now known to be involved in the development of butterfly color patterns. Pigment pathway genes and scale maturation regulators can also have very dramatic effects on the color patterns. Second, population genetics and dynamics models support the prediction that a major phenotypic jump is necessary to cross the deep fitness valleys in a rugged fitness landscape, after which gradual, Fisherian evolution may proceed to enhance resemblance. Finally, experiments show that birds associate cryptic patterns with edibility and generalize those in such a way that only profoundly deviant prey are treated as separate cases by the birds and memorized as warning patterns when appropriate. These experiments also indicate that increased resemblance is still significantly advantageous in imperfect mimics, supporting the second step of the two-step scenario.

Largesse of the Genome

Another, but not exclusive,route to supergenes for mimicry is called largesse of the genome, put forward by J. R. G. Turner. Under this scenario, it is believed that the modification of a trait can be achieved by so many different genes that some of them will inevitably happen to be linked. Among the many possible combinations of loci, selection could simply sieve out the ones that involve linked genes. This hypothesis is particularly likely for loss-of-function phenotypes that can be achieved by mutating any step in the development, like the loss of tail in the African swallowtail P. dardanus. Similarly, that different mimetic species use nonhomolo-gous supergenes can be viewed as indirect evidence for the validity of the largesse of the genome hypothesis in the broad sense.

Supergenes in Mullerian Mimics: A Puzzle

Mullerian mimics being usually monomorphic locally, supergenes are not expected to control wing patterns, and multilocus control was hypothesized to be the norm. This basic prediction has, however, constantly been challenged by Heliconius color-pattern genetics, which show that a limited set of genes of large effect and supergenes control most of the racial color-pattern variation. In the polymorphic H. numata, one single gene seems to control the entire wing pattern, with as many as seven alleles, each allele bringing resemblance to a specific mimetic pattern. Tight gene clusters are also found, to a lesser extent, in polymorphic H. cydno, in H. melpomene, and in H. erato. The existence of these supergenes seems puzzling. It is possible that butterfly color patterns in general are under the control of relatively few conserved genes, at least in some lineages, such as developmental regulatory genes involved in eyespot formation.

In toxic prey, strong selection against any new form and the impossibility of gradual color-pattern changes have been theoretically and empirically demonstrated. It follows that, like Batesian mimics, Mullerian mimics seem to need an initial phenotypic leap, perhaps involving multimodal signal modifications, to jump either to an already protected pattern or away from predators' generalization of cryptic prey. Therefore, it is perhaps no surprise that most exaggerated signal forms studied are under the control of relatively few genes, following the same two-step scenario as in Batesian mimicry. Moreover, switches from one mimetic pattern to another are likely selected only if the new mutant's mimetic characters are not randomly recombined in its descendants. This imposes another constraint (or "sieve") on the genetic architecture for new mimetic patterns to be selected out of a transiently polymorphic population. It is therefore remarkable to note that although Batesian and Mullerian evolutionary dynamics are radically different, and are even perhaps engaged in an evolutionary arms race, the evolution of their signals might require a similar (though nonhomologous) genetic predisposition.

Recent comparative mapping, by M. Joron and colleagues, of the genetic architectures in H. numata and H. melpomene has uncovered a remarkable level of positional genetic homology between those species for the major loci controlling mimicry, despite the fact that H. numata and H. melpomene cover very different mimicry patterns, and are involved in totally disjunct mimicry rings. This suggests that the few genes involved are remarkably flexible in the array of possible phenotypes and that selection for ecological adaptation, as opposed to genetic constraints, is the chief determinant of mimicry diversity.

Myrmecomorphy

Ants represent the most abundant group of organisms in most biota and have powerful multi-modal defenses such as acid taste, aggressive biting, painful sting, and social defense. For these reasons, foraging ants are generally little subject to predation and act as ideal models in mimicry rings. Many insects and spiders indeed have an altered morphology and resemble ants, a phenomenon called myrmecomorphy. For instance, several salticid spider genera such as Myrmarachne or Synemosyna are bewilderingly good ant mimics. It is also common to spot ant-like mirid nymphs (Heteroptera) running among leafcutting Atta ants or Ecitomorpha staphilinid beetles among Eciton army ant columns. The adaptive significance of antlike morphology has been the subject of considerable debate. For instance, several ant-like spiders are believed to mimic ants as a trick to approach and prey on their ant models ("aggressive mimicry"); some ant-like bugs use the same trick to approach and prey on ant-tended aphids. However, most ant-like insects are phytophagous, do not prey on foraging ants, and usually mimic the locally abundant ant species. They are therefore good Batesian mimicry candidates. The interesting aspect of ant mimicry is that, although small birds, lizards, or amphibians may be important predators on ant-sized insects, there are grounds to think that arthropod predators with developed visual skills could be the prime receivers selecting for ant mimicry. For instance, wasps in the Pompilidae are known as important predators of jumping spiders, but ignore ants, thus potentially selecting for ant-like morphology and behavior. Jumping spiders themselves are visual predators hunting insects and also tend to avoid stinging ants as prey. Although the cognitive abilities of arthropods are not well researched, several studies using mantids, assassin bugs Sinea sp., or crab spiders show that they are capable of associative learning and discriminate against ant-like prey. Despite the difference in visual acuity and cognitive abilities between vertebrates and arthropods, it is interesting to note that arthropod predators are likely responsible for visual mimicry that is very accurate to our eyes.

The Importance of Behavior and Motion

Myrmecomorphy highlights a crucial aspect of mimicry: the importance of behavior. Predators integrate many aspects of prey appearance when making a decision of whether to attack, and behavior is an important part of multimodal signals. Ants are characterized by jerky (e.g., Pseudomyrmex spp.) or zigzag (e.g., Crematogaster spp.) movements that their mimics adopt. The rapid movement of antennae is a common feature of ant behavior which mimetic species copy by waving their front legs (e.g., ant-mimicking Salticidae spiders, and spider-wasp-mimicking leaf-footed bugs Coreidae). Because motion considerably enhances visibility, it is hardly surprising that details of the behavior make important identification cues for the predators. For instance, although slow flight in aposematic butterflies may save energy, slowness itself is certainly recognized as such by predators that can select on extremely minute details of flight unnoticeable to the human eye. R. B. Srygley's work on locomotor mimicry has shown that the two butterflies H. erato and H. sapho differ in the asymmetry of the upward and downward wing strokes, which their respective (Mullerian) mimics H. melpomene and H. cydno copy accurately in Panama. Batesian mimics usually retain escape behaviors characteristic of their groups: the lazily flying Neotropical butterfly Consul fabius (Nymphalidae: Charaxinae) can start rapid escape flight when detected; ant-mimicking salticid spiders are also usually reluctant to jump unless attacked.

The tendency for predators to generalize the characteristics of palatable prey, on which they actually feed, probably selects apose-matic signals away from these morphologies, and behavioral signals are no exception. Rapid jerky flight is usually characteristic of a tasty prey, a profit that predators have to weigh against the time and energy costs associated with catching the prey. Unconventional behaviors like the flight of Heliconius butterflies or the looping of honey bees make them highly noticeable to predators. This imposes an additional visibility cost on incipient mimetic prey; for the resemblance to be selected, such cost has to be offset by a significant reduction in predation. These considerations suggest that mimetic behavioral change probably evolves in much the same way as morphological characters do, that is, a two-step process.

Escape Mimicry

Unpalatability is not the only way to be unprofitable to predators. Fast, efficient escape is another way for preys to teach predators that pursuit is useless and will bring no reward: predators unable to consume the desired prey may associate this frustration with the prey appearance and reduce their attacks on this prey altogether. Even if the prey can be seized, predators probably trade off the energy spent and the (often low) nutritional reward. In several experiments birds were shown to be able to decrease their attack rates when the presented prey would quickly disappear ("escape") during their attacks, and conspicuousness of the prey tended to enhance the response. Therefore, evasive prey could advertise their escaping abilities by color patterns, which other prey may mimic. At least three kinds of characters may enhance the difficulty of catching an evasive prey: erratic flight (like that of pierids), fast and maneuver-able flight (like that of charaxine butterflies), or high reactivity (like that of syrphid hoverflies). Typically, these escape specialists are all palatable to predators. Some species of the Neotropical butterfly genera Adelpha (subfamily Nymphalinae) and Doxocopa (subfamily Charaxinae) show convergent appearance and exhibit extremely quick escape when slightly disturbed, followed by very fast flight. Their resemblance is hypothesized by R. B. Srygley to be a case of escape mimicry. The poor resemblance of some hoverflies to their purported hymenopteran models has also led to the hypothesis that groups of syrphid species could represent an escape mimicry ring on their own.

Poor Mimicry

At least to our eyes, the model's color pattern is not always copied very accurately. Many syrphid flies, for instance, are difficult to assign to particular mimicry rings, although they seem to mimic the general appearance of Hymenoptera. The heterogeneity in mimetic accuracy has led biologists to propose adaptive and nonadaptive hypotheses, none of which seems very strongly supported at present. (1) The null hypothesis is that poor mimics are no mimics: many mimicry associations have been claimed on the general appearance of an insect, whereas careful examination of the geographic covariation of purported models and mimics may reveal evidence against them. In the case of inaccurate mimics, this method is not very powerful because the mimetic association itself is hard to define; so such covariation is difficult if not impossible to judge. (2) Another nonadaptive scenario is that accurate mimicry may not always be possible, either because of functional constraints/trade-offs on the modified organs or because of genetic or developmental constraints on the variation available in populations. Mimicry may then asymptotically reach a maximum level of resemblance, contingent on the route followed in the initial stages of the mimetic change. Again, this is theoretically plausible, but difficult to test.

(3) Among the adaptive explanations for inaccurate mimicry is the hypothesis that these species are in the initial stages of their mimetic change and that our instantaneous view of evolution doesn't show us the complete picture. (4) Another adaptive scenario is that predators have biases and perceptions different from those of humans and are likely to generalize more in some directions than in others, leading to the possibility that mimics that look very inaccurate to us are in fact very good mimics for a predator. Generalization is also dependent on the strength of the harmfulness of the models, perhaps allowing lower levels of accuracy. This may be the case for poor mimicry in some hoverflies. The ultimate adap-tationist hypothesis is that inaccuracy itself may be beneficial. (5) It could either allow the mimic to benefit from the protection of several different models, perhaps in a heterogeneous environmental context, or—a related hypothesis—create conflict in the predators' recognition, which may give the mimic more time and chances to escape.

Mimicry, Community Ecology and Macroevolutionary Patterns

Habitat Heterogeneity, Spatial Dynamics, and the Coexistence of Mimicry Rings. The efficiency of a warning pattern depends on the abundance of that pattern in the habitat. Therefore, as new species join a particular mimicry ring, the protection given by the pattern increases, and more species should converge on this best protected pattern.

Ultimately, all species should converge on a single mimicry ring. But nature seems to behave in a totally different way. In any one habitat, particularly in tropical environments, aposematic insects of similar size and shape usually cluster into a number of distinct mimicry complexes or mimicry rings.

Multiple Mimicry Rings in the Community

One possibility is that different mimicry rings are found in different microhabitats. If predators do not move between microhabitats, or retain microhabitat-specific information, insect species in different microhabitats could converge on different adaptive peaks. Flight height has been invoked as a possible explanation, following the rainforest stratification paradigm, but evidence from butterflies is rather equivocal. However, host-plant stratification and different nocturnal roosting heights in Neotropical butterflies have received empirical support. Forest maturity and succession stage influence the host-plant composition and may allow the maintenance of multiple mimicry rings in a mosaic habitat.

Multiple Mimicry Rings within a Species

If some species are patchily distributed because of their microhabitat requirements, each "sub-population" may be particularly sensitive to genetic drift and allow the local predators to learn and select a different color pattern in different patches. Once locally stabilized, the new pattern may be hard to remove. Indeed, local positive frequency dependence is both very efficient at stabilizing patterns around fitness peaks and slow at removing already established suboptimal patterns. Any slight difference in microhabitat quality or patchiness of the species involved will increase the local apparent abundance of particular patterns to particular predators, further decreasing the power of selection to achieve ultimate convergence.

This "mosaic mimetic environment" theory can help explain some problematic cases of Mullerian polymorphism. For instance, Laparus doris is a Heliconiine butterfly (Nymphalidae) that has up to four coexisting forms in some populations, some of which are probably mimetic and others are not. The maintenance of polymorphism in this species could be attributed to its high larval and pupal gregari-ousness (several hundreds of individuals), which results in a patchy distribution of the adults. When hundreds of butterflies suddenly emerge from one single vine, they make up their own local mimetic environment, and the mimetic environment prior to the mass emergence might be effectively neutral to L. doris.

If the species composition and the resulting mimetic environment are spatially variable, polymorphism can evolve in microhabitat gen-eralists, with gene flow across these microhabitats. For example, the Amazonian polymorphic species H. numata is selected toward different mimetic patterns in different localities that may represent different microhabitats for their more specialized models in the genus Melinaea (subfamily Ithomiinae). The balance between local selection and gene flow in a mosaic habitat (and perhaps weak selection against polymorphism as suggested earlier) can therefore maintain a nonadaptive, although widespread, polymorphism in H. numata.

Coevolution in Mimicry

Evolutionary Rates and the Coevolutionary

CHASE Despite many potential sieves constraining mimicry, several to many edible species can end up mimicking a particular warning pattern in a parasitic way. In such cases, is it possible that a "Batesian-overload" threshold is reached, beyond which the efficiency of the signal is severely lowered? Batesian mimics are indeed parasites of the honest signals of their models, and so the models should escape their mimics by evolving a new warning pattern. However, this escape would be transient because the new pattern would soon attract new Batesian mimics, resulting in an evolutionary arms race, or coevolutionary chase, between the model and its mimics. Some authors suggested that this chase could be a cause of the mimetic diversity in both models and mimics and that cyclical interactions could arise in some cases. However, first, theory has shown that mimics always evolve faster than their models, because they gain a lot more from mimicry than models lose from being mimicked. Any gradual move of the model should be quickly matched by a similar evolution in the mimic. Second, the models, which are the prime educators of local predators, are under strong purifying selection against any new warning pattern. This strong intraspecific conservative force should in the vast majority of cases be stronger than the deleterious effects of being mimicked and preclude pattern change in the models. Coevolutionary changes between Batesian mimics and their models should therefore be stopped in their early stages by a stronger selection for the status quo, and both the models and their mimics should be trapped in the same warning pattern. Only by a phenotypic leap toward an already established warning pattern (Mullerian mimicry) or by crossing a fitness valley thanks to local genetic drift could the model ever escape its mimics.

Mutualism and Coevolution in Mullerian

MIMICRY In contrast with the unilateral Batesian evolution in which mimics outrun their models, Mullerian mimicry was traditionally thought to involve mutual resemblance of the species involved, as if all had moved toward some halfway phenotype. Of course, Muller himself and others were quick to point out that the mutual benefits were not even, but lopsided, that is, typically the rarer or the less distasteful species would benefit more than the more common or better defended one (respectively). However mutualistic the relation is, coevolution has often been assumed in Mullerian associations, and the protagonists are usually called comimics just because it is difficult to know if one species is driving the association. Coevolution also predicts that geographic divergence and pattern changes should be parallel in both species of comimics, like in the mimetic pair H. erato and H. melpomene in tropical America, presumably leading to parallel phylogenies. However, DNA sequences from mito-chondrial and nuclear genes show distinct phylogenetic topologies in these two species and distinctly nonparallel evolution.

In fact, there are a number of grounds on which to believe that the asymmetrical relationship leads to one-sided signal evolution even in Mullerian mimicry, one species being a mimic and the other a model. First, because of number dependence, mimetic change of a rarer species toward a commoner species will be retained, but the reverse is not true: by mimicry of a less common species, the commoner species would lose the protection of its own ancestral pattern, and a change toward a rarer pattern would be initially disadvantageous. The commoner species is therefore effectively locked in its pattern, and initial changes are only likely in the rarer species. Second, given the selection against nonmimetic intermediates, the mutants in the rarer species will have to be roughly mimetic of their new model to be selected, thus bringing the ultimate shared signal closer to that of the common species. Once this initial step is made by the mimic, there could be gradual "coevolution" to refine the resemblance, but the resulting change in color pattern will inevitably be more pronounced in the mimic, the model remaining more or less unchanged. Because Mullerian pairs are of a mimic-model nature, even with mutual benefits, the prediction for parallel evolution is therefore not likely to be valid. Indeed, in the mimetic pair H. erato/H. melpomene, the phylogeography suggests that H. melpomene has radiated onto preexisting H. erato color-pattern races, thus colonizing all color-pattern niches protected by H. erato in South America.

Mimicry, Speciation and Radiations

Racial boundaries in mimetic butterflies are usually very permeable to genetic exchange, since selection acts primarily on color-pattern genes. However, because clines moving geographically are likely stopped at ecological boundaries, the resulting racial boundaries are likely to rest on ecological gradients. Racial boundaries between mimetic color patterns could therefore be reinforced by adaptation to local ecology on either side of the cline, leading to spe-ciation. Color-pattern diversification could then accelerate specia-tion by allowing both postmating reproductive isolation, because of a higher mortality of nonmimetic hybrid offspring, and premating isolation if color pattern itself is used as a mating cue by the insect. For these reasons, mimicry has the potential to accelerate specia-tion. The pattern of mimetic associations in Heliconius butterflies seems indeed to indicate that speciation and mimetic switches are usually coincident: sister species usually differ in their mimetic color pattern. Direct evidence of the role of color pattern in mate choice has been gathered for the sister species pairs H. erato/H. himera and H. melpomene/H. cydno . The first two species are geographically separated across an ecological gradient in the Andes. The second pair is sympatric, although the species also differ in ecological requirements in a patchy distribution. In both pairs, therefore, color-pattern and mimetic switches probably accelerated speciation initiated by ecological adaptation. It is unknown how general this mimicry-based speciation is in mimetic insects but it could be an important consequence of the rampant and apparently easy diversification of mimetic patterns at the intraspecific level. The genetic predisposition of mimetic species to evolve polymorphism—the first stage toward speciation—might explain why mimetic lineages are usually very speciose and undergo rapid radiations, both geographically and phylogenetically.

Chemical Defense

Biologists have become keenly aware that insects possess a remarkable ability to biosynthesize a large variety of compounds for use as agents of chemical defense against their omnipresent enemies. Many of these compounds are unique products (e.g., cantharidin, or Spanish fly, produced by blister beetles) with diverse modes of toxicity against a variety of vertebrate and invertebrate predators. These defensive secretions often originate from unlikely sources that appear to optimize the effectiveness of the chemical defensive systems. Ultimately, for countless species of insects, chemical defense and survival are synonymous.

Eclectic Origins, Functions and Reservoirs of Defensive Compounds

It would be no exaggeration to state that the tremendous abundance of insects constitutes the primary food source for diverse vertebrate and invertebrate predators. For insects in a variety of orders, blunting the attacks of their omnipresent predators is identified either with the production of defensive compounds in exocrine glands or with the acquisition of these compounds from external sources. These deterrent allomones sometimes represent novel natural products that have a very limited distribution in the Insecta. In short, exocrine compounds, characteristic of species in orders or genera, have evolved to function as versatile agents of chemical defense.

It has been generally assumed that de novo biosynthesis characterizes the origins of insect defensive compounds. However, recent investigations suggest that novel insect defensive allomones, including the complex amide pederin from staphylinid beetles (Paederus spp.), and unique steroids from dytiscid beetles, are biosynthesized by endosymbiotes. These results raise the question of whether other novel insect allomones, including the terpene cantharidin, and steroids in chrysomelids and lampyrids, may have microbial origins.

Often, however, the deterrent allomones constitute ingested alle-lochemicals such as cardenolides (milkweeds) and toxic pyrrolizidine alkaloids (asters, heliotrope). Furthermore, some of these plant natural products have been metabolized after ingestion into products that are suitable for sequestration and use as deterrents, as is the case for ingested steroids from milkweeds by the monarch butterfly, Danaus plexippus. These compounds are also transferred to eggs to function as effective predator deterrents. In addition, these allelochemicals may be added to the secretions of exocrine glands, further increasing the deterrent properties of these exudates. The dependence on ingested plant natural products of some insect species is further emphasized by the utilization of " stolen " defensive exudates that essentially represent mixtures of pure allelochemicals that have been appropriated, unchanged, from their host plants.

In some species, ingested allelochemicals are sexually transmitted by the male as a copulatory "bonus" for the female. For example, the sperm-rich spermatophore of ithomiine butterflies is accompanied by pyrrolizidine alkaloids that provide protection for the female and her eggs. Importantly, this very adaptive system is functional because the spermatozoa are resistant to the well-known toxic effect of these alkaloids.

Some allelochemicals also possess great selective value for insects as antibiotic agents. Alkaloids such as alpha-tomatine, a constituent of tomatoes, reduce the infectivity of bacteria and fungi for lepidop-terous larvae. Other compounds reduce the activity of viruses and in some cases are highly toxic to insect parasitoids.

The defensive value of insect allomones has been further enhanced by the ability of these arthropods to adapt a variety of these natural products to subserve a surprising variety of multiple functions. This phenomenon, semiochemical parsimony, has been particularly emphasized by insect species such as fire ants, whose alka-loidal venoms possess a dazzling variety of pharmacological activities. The same may be said of cantharidin (Spanish fly), the potent vesicant from blister beetles.

Things are seldom what they seem. The sting-associated glands of bees and wasps are obvious candidates for the production of compounds with considerable deterrent activities. These glands have evolved as biosynthetic centers clearly dedicated to the biogenesis of pharmacologically active compounds that can be delivered by the sting in an unambiguous act of defense. On the other hand, some glands clearly identified with nondefensive functions have been adapted by a variety of insect species to function as defensive organs with varied functions. Furthermore, the deterrent efficiency of these secretions may be considerably enhanced by adding repellent plant natural products to the exudates. And insects have not neglected adapting enteric products to discourage their omnipresent predators. If all else fails, many insects eject blood, sometimes fortified with toxic allomones, at their adversaries, with startling results. It is no exaggeration to state for these species, bleeding has often provided an extraordinary means of deterring a variety of aggressive predators.

Variety of Salivary Defensive Functions

Salivary Venoms

The spitting cobra, Naja nigricollis. has an insect parallel, both in terms of the general chemistry of the saliva and the ability to accurately "fire" the venom at a moving target. For example, Platymeris rhada-manthus is a black and orange assassin bug (Reduviidae) that is very conspicuous because of its aposematic (warning) coloration. This insect can eject its saliva for a distance of up to 30 cm, and if this enzyme-rich solution (proteases, hyaluronidase, phospholipase) strikes the nose or eye membranes of a vertebrate, intense pain, edema, and considerable vasodilation may follow. The saliva of P. rhadamanthus is admirably suited to deter vertebrate predators including birds and reptiles. This rapidly acting salivary venom has clearly evolved for predation on invertebrates.

Entspannungschwimmen (Chemically Induced Aquatic Propulsion)

The proteinaceous saliva of the hemipteran Vela capraii has been adapted to promote escape from potential predators in aquatic environments. This aquatic true bug will discharge its saliva onto the water surface, a reaction that results in lowering the surface tension of the water behind the bug and propelling it across the aquatic surface.

Allomonal Pheromones

Many bumble bees (Bombus spp.) scent mark territorial sites with cephalic products that are very odoriferous. The secretions, which originate in the cephalic lobes of the salivary glands, are dominated by terpenes, some of which are well-known defensive compounds. This appears to be an excellent example of semiochemical parsimony, with the males utilizing the compounds both as territorial pheromones and as defensive allomones.

Salivary Glues

Termite workers in both primitive and highly evolved genera secrete defensive exudates that are rapidly converted to rubberlike or resinous products that can rapidly entangle small predators such as ants. This conversion frequently reflects the polymerization of salivary proteins that have reacted with p-benzoquinone, a highly reactive defensive product. Similar systems for generating entangling salivas have been detected in a diversity of termite genera, including Mastotermes, Microtermes, Hypotermes, and Odontotermes.

Termites in other genera discharge cephalic exudates that are fortified with toxic terpenes. Species of Nasutitermes and Tenuirostritermes secrete mixtures of compounds that rapidly form a resin that entangles ants and other small predators. The presence of monoterpene hydrocarbons is probably responsible for killing ants, and may function as an alarm pheromone for recruiting termite soldiers.

Nonsalivary Entangling Secretions

The posterior abdominal tergites and cerci of cockroaches in a variety of genera are covered with a viscous secretion that can act as an entangling glue for small predators. Species in genera as diverse as Blatta and Pseudoderopeltis produce proteinaceous secretions on the abdominal tergites that would be readily encountered by predators pursuing these cockroaches. After seizing the cockroaches, centipedes, beetles, and ants rapidly release their prey while cleaning their mouthparts. The fleeing cockroaches generally have more than ample time to affect their escape.

Aphid species in many genera also utilize an entangling secretion as a primary means of defense. The exudate is discharged in response to a confrontation, often hardening to a waxy plaque on an adversary within 30 s. This defensive behavior, which appears to be widespread in the Aphididae, uses tubular secretory organs, the cornicles, on the fifth and sixth abdominal tergites. The secretions, which are dominated by triglycerides, have been characterized in a range of genera, including Aphis, Myzus, Acyrthosiphon, and Therecaphis. The cornicular secretions are clearly more effective against generalized predators (e.g., ants) than they are against specialized predators (coccinellids, nabids). The secretions also contain alarm pheromones, E-B-farnesene and germacrene A, which release dispersive behavior that may cause aphids to drop off plants.

A variety of glands have been evolved by ants as sources of viscous defensive secretions. Many species in the subfamily Dolichoderinae discharge a pygidial gland secretion that is dominated by cyclopenta-noid monoterpenes such as iridodial, compounds that rapidly polymerize on exposure to air. The viscous polymer effectively entangles small predators such as ants. Myrmicine species in the genus Pheidole also use the pygidial glands as a source of entangling glue and in addition, an alarm pheromone.

Defensive Froths from Diverse Glands

A surprising diversity of defensive secretions has been converted to froths that may literally bathe small adversaries with compounds that seem to adversely stimulate the olfactory and gustatory receptors of their predators. The independent evolution of deterrent froths by moths, grasshoppers, and ants demonstrates that this form of defensive discharge can be highly efficacious in adverserial contexts.

Species in the moth families Arctiidae (aposematic tiger moths), Hypsidae, and Zygaenidae often secrete froths, the production of which may be accompanied by a hissing sound and a pungent odor. The aposematism of these moths is enhanced by secretions discharged from brightly colored areas near or on the prothorax. These secretions do not seem to contain plant natural products but rather, toxic de novo synthesized compounds such as pharmacologically active choline esters.

Frothing is highly adaptive in the ant genus Crematogaster. Workers in this very successful myrmicine genus do not possess a hypodermic penetrating sting, but rather, a spatulate sting that is enlarged at the tip. Venom accumulates at the tip and can be smeared onto small adversaries such as ants, as if with a paintbrush. This mode of administration of venom is obviously identified with a topical toxicant that can penetrate the insect cuticle much as an insecticide does.

Two grasshopper species produce froths that are derived from a mixture of tracheal air and glandular secretion. Both species are eminently aposematic, and this warning coloration is enhanced by a powerful odor emanating from the plant-rich froths of the pyrgo-morphid Poekilocerus bufonius, a specialist milkweed feeder, and the acridid Romalea guttata, a generalist feeder with very catholic tastes.

Externalizing Allomones by Reflex Bleeding

Many insect species, particularly beetles, externalize their distinctive defensive compounds in a blood carrier rather than discharging them as components in a exocrine secretion.

Cantharidin, the terpenoid anhydride synthesized by beetles in the families Meloidae and Oedemeridae, is externalized in blood discharged reflexively from the femorotibial joints. The repellent properties of cantharidin were established more than 100 years ago, and the ability of amphibians to feed on these toxic beetles, with impunity, has been long known, as well. Cantharidin possesses a wide spectrum of activities, including inducing priapism in the human male, and it has been reported to cause remission of epidermal cancer in mammals. Although its role as a repellent and lesion producer certainly documents its efficacy as a predator deterrent, its potent antifungal activity may protect developing meloid embryos from entomopathogenic fungi present in their moist environment.

Autohemorrhage from the femorotibial joints is widespread in many species of ladybird beetles (Coccinellidae), most of which are apose-matic. The blood is generally fortified with novel alkaloids that are outstanding repellents and emetics (i.e., inducers of vomiting) as well.

Adult fireflies (Photinus spp.) produce novel steroids (lucibufa-gins) that are effective repellents and inducers of emesis in invertebrates and vertebrates. Reflex bleeding from specialized weak spots in the cuticle along the elytra and antennal sockets externalizes these steroids. Sometimes rapidly coagulating blood, free of allomones, is used defensively.

Blood as Part of a Glandular Secretion

Often the secretions of defensive glands are fortified with blood. For example, arctiid moths (e.g., Arctia caja) discharge odoriferous froths from prothoracic glands, and these exudates contain pharmacologically active choline esters that are accompanied by blood. A similar system characterizes the pyrgomorphid grasshopper P. bufonius.

Nonglandular Discharges of Plant Origin

Certain insects have evolved storage reservoirs for plant natural products that can be discharged in response to traumatic stimuli. The evolutionary development reflects the insect's appropriation of plant allelochemicals (defensive compounds) for subsequent utilization as defensive allomones. In essence, the insects have sequestered the plant's defenses and stored them in reservoirs where they are available as defensive agents. This defensive system does not require the evolution of any biosynthetic pathways for the synthesis of compounds stored in nonglandular reservoirs.

Adults of hemipterous species in the family Lygaeidae possess dorsolateral (reservoirs) and abdominal spaces that contain a fluid very similar to that of the proteins in the blood. This fluid sequesters steroids (cardenolides) present in the milkweeds on which these species feed. The cardenolides are about 100-fold more concentrated in the dorsolateral fluid than they are in the blood, and they constitute a formidable deterrent system.

Sequestration of plant natural products in nonglandular reservoirs also characterizes larvae of the European sawfly, Neodiprion sertifer. Feeding on pine (Pinus spp.), these larvae sequester both deterrent mono- and sesquiterpenes in capacious diverticular pouches of the foregut.

Plant Natural Products in Exocrine Secretions

Herbivorous insects may incorporate plant natural products into exocrine and nonexocrine defensive secretions. By selectively adding plant repellent compounds to their own deterrent secretions, insects can increase the effectiveness of their own chemical deterrents. These plant-derived compounds are generally unrelated to the constituents in the exudates of their herbivores. In all likelihood, these plant additives may augment the repellency of the deterrents by reacting with olfactory chemoreceptors different from those targeted by the insect-derived repellents.

The large milkweed bug, Oncopeltus fasciatus, in common with many species of true bugs, uses the secretion of the metathoracic scent gland as an effective defensive exudate. Nymphs of this species generate defensive secretions with middorsal glandular fluid. The repellent secretions also contain cardenolides derived from the milkweed host plants of this species.

Similarly, the acridid R. guttata sequesters, in the metathoracic defensive glands, plant allelochemicals that can considerably augment the deterrent effectiveness of the secretion. Unlike O. fascia-tus, R. guttata is a generalist that feeds on and sequesters a potpourri of plant natural products. As a consequence, the compositions of the glandular exudates can be variable, sometimes resulting in secretions that are considerably more repellent than those derived from insects that had fed on a limited number of host plant species.

Regurgitation and Defecation of Allelochemicals

Enteric defense may be widespread in insects as a means of using the proven repellencies of a variety of plant natural products. In a sense, the intestine is functioning as a defensive organ once repellent plant products have been ingested, and it is likely that the presence of pharmacologically active plant compounds in the intestine renders the insect distasteful or emetic. Therefore, transfer of gut contents to the outside either by regurgitation, or by defecation, as seen in milkweed bugs (Oncopeltus spp.), could actually constitute the externalization of the internal enteric defenses.

Lubber grasshoppers readily regurgitate when subjected to traumatic stimuli and the plant-fortified discharge readily repels ants. The same is true of molested larvae of the lymantriid moth Eloria noyesi which produce a discharge containing cocaine from their host plant.

Mechanical Defenses

Morphological structures of predatory function, such as the modified mouthparts and spiny legs also may be defensive, especially if a fight ensues. Cuticular horns and spines may be used in deterrence of a predator or in combating rivals for mating, territory, or resources, as in Onthophagus dung beetles. For ectoparasitic insects, which are vulnerable to the actions of the host, body shape and sclerotization provide one line of defense. Fleas are laterally compressed, making these insects difficult to dislodge from host hairs. Biting lice are flattened dorsoventrally, and are narrow and elongate, allowing them to fit between the veins of feathers, secure from preening by the host bird. Furthermore, many ectoparasites have resistant bodies, and the heavily sclerotized cuticle of certain beetles must act as a mechanical antipredator device.

Many insects construct retreats that can deter a predator that fails to recognize the structure as containing anything edible or that is unwilling to eat inorganic material. The cases of caddisfly larvae (Trichoptera), constructed of sand grains, stones, or organic fragments, may have originated in response to the physical environment of flowing water, but certainly have a defensive role. Similarly, a portable case of vegetable material bound with silk is constructed by the terrestrial larvae of bagworms (Lepidoptera: Psychidae). In both caddisflies and psychids, the case serves to protect during pupation. Certain insects construct artificial shields; for example, the larvae of certain chrysomelid beetles decorate themselves with their feces. The larvae of certain lacewings and reduviid bugs cover themselves with lichens and detritus and the sucked-out carcasses of their insect prey, which can act as barriers to a predator, and also may disguise themselves from prey.

The waxes and powders secreted by many hemipterans (such as scale insects, woolly aphids, whiteflies, and fulgorids) may function to entangle the mouthparts of a potential arthropod predator, but also may have a waterproofing role. The larvae of many ladybird beetles (Coccinellidae) are coated with white wax, thus resembling their mealybug prey. This may be a disguise to protect them from ants that tend the mealybugs.

Body structures themselves, such as the scales of moths, caddisflies, and thrips, can protect as they detach readily to allow the escape of a slightly denuded insect from the jaws of a predator, or from the sticky threads of spiders' webs or the glandular leaves of insectivorous plants such as the sundews. A mechanical defense that seems at first to be maladaptive is autotomy, the shedding of limbs, as demonstrated by stick-insects (Phasmatodea) and perhaps crane flies (Diptera: Tipulidae). The upper part of the phasmatid leg has the trochanter and femur fused, with no muscles running across the joint. A special muscle breaks the leg at a weakened zone in response to a predator grasping the leg. Immature stick-insects and mantids can regenerate lost limbs at molting, and even certain autotomized adults can induce an adult molt at which the limb can regenerate.

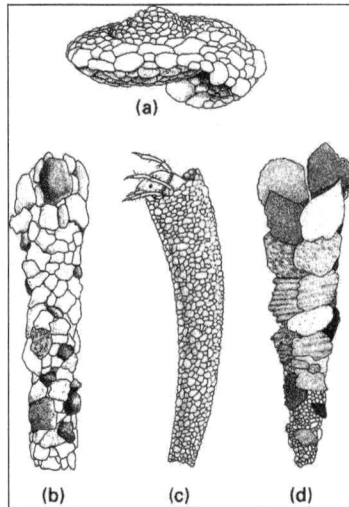

Portable larval cases of representative families of caddisflies (Trichoptera):
(a) Helicopsychidae; (b) Philorheithridae; (c) and (d) Leptoceridae.

Secretions of insects can have a mechanical defensive role, acting as a glue or slime that ensnares predators or parasitoids. Certain cockroaches have a permanent slimy coat on the abdomen that confers protection. Lipid secretions from the cornicles (also called siphunculi) of aphids may gum-up predator mouthparts or small parasitic wasps. Termite soldiers have a variety of secretions available to them in the form of cephalic glandular products, including terpenes that dry on exposure to air to form a resin. In Nasutitermes (Termitidae) the secretion is ejected via the nozzle-like nasus (a pointed snout or rostrum) as a quick-drying fine thread that impairs the movements of a predator such as an ant. This defense counters arthropod predators but is unlikely to deter vertebrates. Mechanical-acting chemicals are only a small selection of the total insect armory that can be mobilized for chemical warfare.

References

- Defense: cals.ncsu.edu, Retrieved 13 May, 2019

- Amazing-insect-defensive-tactics: ranker.com, Retrieved 3 April, 2019

- Mimicry-insects, insects: what-when-how.com, Retrieved 9 July, 2019

- Chemical-defense-insects, insects: what-when-how.com, Retrieved 19 August, 2019

Permissions

All chapters in this book are published with permission under the Creative Commons Attribution Share Alike License or equivalent. Every chapter published in this book has been scrutinized by our experts. Their significance has been extensively debated. The topics covered herein carry significant information for a comprehensive understanding. They may even be implemented as practical applications or may be referred to as a beginning point for further studies.

We would like to thank the editorial team for lending their expertise to make the book truly unique. They have played a crucial role in the development of this book. Without their invaluable contributions this book wouldn't have been possible. They have made vital efforts to compile up to date information on the varied aspects of this subject to make this book a valuable addition to the collection of many professionals and students.

This book was conceptualized with the vision of imparting up-to-date and integrated information in this field. To ensure the same, a matchless editorial board was set up. Every individual on the board went through rigorous rounds of assessment to prove their worth. After which they invested a large part of their time researching and compiling the most relevant data for our readers.

The editorial board has been involved in producing this book since its inception. They have spent rigorous hours researching and exploring the diverse topics which have resulted in the successful publishing of this book. They have passed on their knowledge of decades through this book. To expedite this challenging task, the publisher supported the team at every step. A small team of assistant editors was also appointed to further simplify the editing procedure and attain best results for the readers.

Apart from the editorial board, the designing team has also invested a significant amount of their time in understanding the subject and creating the most relevant covers. They scrutinized every image to scout for the most suitable representation of the subject and create an appropriate cover for the book.

The publishing team has been an ardent support to the editorial, designing and production team. Their endless efforts to recruit the best for this project, has resulted in the accomplishment of this book. They are a veteran in the field of academics and their pool of knowledge is as vast as their experience in printing. Their expertise and guidance has proved useful at every step. Their uncompromising quality standards have made this book an exceptional effort. Their encouragement from time to time has been an inspiration for everyone.

The publisher and the editorial board hope that this book will prove to be a valuable piece of knowledge for students, practitioners and scholars across the globe.

Index